W9-BXW-486

3 1215 00098 4226

# Exploring Biomechanics

*Animals in Motion*

# EXPLORING BIOMECHANICS

*Animals in Motion*

R. McNeill Alexander

SCIENTIFIC
AMERICAN
LIBRARY

A Division of HPHLP
New York

Library of Congress Cataloging-in-Publication Data

Alexander, R. McNeill.
Exploring biomechanics: Animals in motion / R. McNeill Alexander.
p. cm.
Includes bibliographical references (p. ) and index.
ISBN 0-7167-5035-X
1. Animal locomotion I. Title.
QP301.A2958 1992
591.1'852—dc20 91-37293
CIP

ISSN 1040-3213



Printed in the United States of America.

Scientific American Library
A Division of HPHLP
New York

Distributed by W. H. Freeman and Company.
41 Madison Avenue, New York, New York 10010 and
20 Beaumont Street, Oxford OX1 2NQ, England

2 3 4 5 6 7 8 9 0 KP 9 9 8 7 6 5 4 3 2

This book is number 40 of a series.

# Contents

# Preface

Biology and mechanical engineering may seem to be widely separate branches of knowledge, but biomechanics is the science that brings them together. Biomechanics takes the understanding of structures, machines, and vehicles that engineers have developed and uses it to explain how living things move and work. One of the liveliest and most fascinating branches of biomechanics is the study of animal locomotion: how birds fly and fishes swim, how slugs crawl, how legged animals (and people) run, and much else besides. The bewildering variety of styles of movement used by creatures ranging in size from amoebas to whales presents innumerable challenges to biomechanicists.

This book tells in words and shows in pictures how animals move. It describes the many ingenious experiments and illuminating ideas that have given us our present understanding of animal movement, and it points to some of the problems that remain to be solved.

R. McNeill Alexander
December 1991

# 1

# Muscle: The Motor for Animal Movement

A Red fox (*Vulpes vulpes*) pounces on a mouse. At take-off, the muscles of the hind legs throw the animal into the air: most of this work is done by the big muscles of the thighs. On landing, the muscles of the fore legs function as brakes, halting the animal's fall.

Some animal movements are intensely beautiful: think, for example, of the soaring of eagles and the running of gazelles. Others are very puzzling. How does an insect walk upside down on the ceiling or the underside of a leaf? How do amoebas crawl? Yet other movements may seem easy to explain if you know only a little but become profoundly puzzling when you learn more.

This book is about all forms of animal locomotion: walking, running, and jumping; crawling and climbing; swimming and flight. In addition to animal movement, it discusses human movement and the movements of single-celled organisms like amoebas. It explains (as far as our current knowledge allows) how animals move and how their bodily structure and their patterns of movement are adapted to save energy and maximize performance. I have tried to explain the basic principles of animal locomotion and to present the results of the latest and most enlightening research.

One aspect of this book may surprise you. You will expect to find biological concepts in a book on animal movement, but you may be less prepared for ideas from engineering and physical science. Yet you will find physical ideas cropping up on almost every page, for a very good reason. We want to understand how animals move and how evolution has adjusted their structure and movements to maximize performance. If you wanted to understand how a motor vehicle moved and how it was designed for high performance, you would apply your knowledge of engineering and physical science. The same physical laws apply to the movement of living things as to the working of machinery, so we need physical ideas to understand the move-

This caterpillar crawls by a looping action, gripping the twig alternately with the true legs at the front end of its body and the false legs at the rear. With its hind end attached, it reaches forward as far as possible to attach the front end. Then it draws up the hind end close to the front, reattaches it, and repeats the process.

ments of animals, and many other biological problems. Engineering, physics, and biology are far less separate from each other in modern science than they often seem to be in the school curriculum.

A good way to start, if we were studying motor vehicles, would be to learn about the engines that drive them. Similarly, in this book about animal movement, it seems sensible to start with the muscles that power the movements of most animals. We will soon find ourselves using such physical concepts as force, work, and energy.

## Exerting Force with Muscles

We are familiar with muscle as the meat in beef, pork, chicken, and fish. There is more of this tissue than of any other in the bodies of many animals: for example, muscle accounts for 63 percent of the body weight of trout and 46 percent of the body weight of the Uganda kob antelope. Muscle is responsible for driving the body's movements. It powers the galloping of horses, the flight of bees, and even

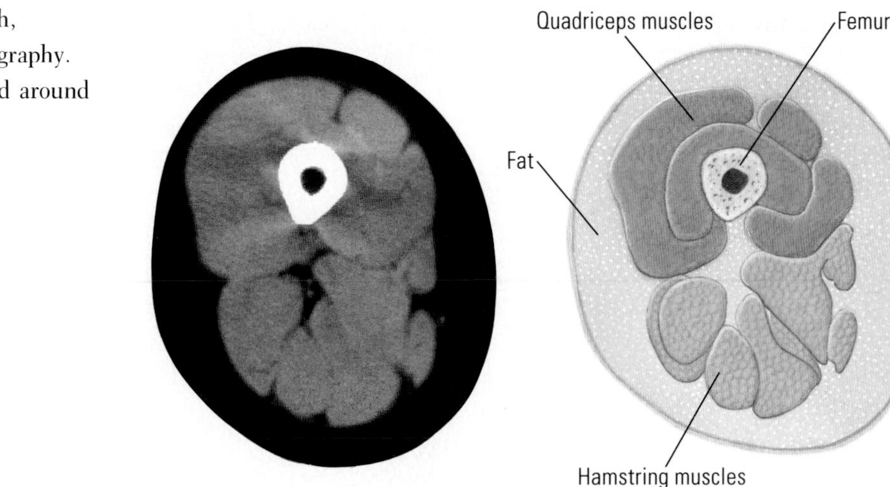


A section of a living human thigh, revealed by computer-aided tomography. The muscles can be seen grouped around the central bone.

the swimming of jellyfishes. Only in very small animals such as rotifers and other tiny plankton are body movements driven principally by motors other than muscle. These motors are described near the end of this book, in Chapter 8.

To power movement, muscles must exert forces: forces to support the animal's weight, forces to accelerate or decelerate the animal, and forces to overcome the resistance of the air or water through which the animal moves. These forces are transmitted to the environment through structures such as limbs and fins. The muscles of our legs, for example, exert forces on the skeleton, making our feet push on the ground, and the muscles of a fish's tail exert forces that make the tail fin push on the water. We want to know how these forces are produced.

Much of what we know about how muscles work and about the forces they can exert comes from experiments on muscles or parts of muscles from the legs of small animals such as frogs and rats. If these are removed from the animal's body very soon after death and placed in solutions that resemble its body fluids, they can be kept alive for many hours. During this time they can be made to contract by electrical stimulation, and their properties can be investigated.

These experiments have confirmed what we might have expected, that stout muscles are stronger than slender ones: the force that a muscle can exert depends on its cross-sectional area. To take account of this, physiologists often divide the force by the cross-sectional area. The result of this simple calculation is the stress that a muscle can exert. Experiments on muscles from frogs, rats, and other vertebrates show that however stout or slender these muscles are, all can exert up to about 0.3 newton (that is, 30 grams force) per square millimeter of muscle section—44 pounds force per square inch. This is not a very large stress (it is approximately the same as the stress in a rubber band stretched by 30 percent of its original length), but muscles with large cross-sectional areas can exert very large forces. For example, the cross-sectional areas of muscles in the legs of living people can be measured by computer-aided tomography, an X-ray technique often known as a CAT scan that produces pictures representing sections through the body. Measurements of the hamstring muscles at the back of the thighs of healthy young men show that they have a total cross-sectional area of about 15,000 square millimeters, enough to exert 4500 newtons, or 1000 pounds force. More slender muscles exert proportionately smaller forces, but even these can be quite substantial; for example, the small muscle that bends the human thumb can exert a force of 130 newtons (30 pounds force).

Like other living tissues, muscle consists of cells, but its cells have an unusual shape: because they are very long and slender, they are called muscle fibers. A cat muscle fiber only 40 micrometers (0.4 millimeter) in diameter may be as much as 40 millimeters (1.6 inches) long. The relatively larger size of these fibers has consequences for their internal structure. Whereas most other cells have just one nucleus each, the genes in a single nucleus could not exert effective control over the metabolism of a cell with the length of a muscle fiber. This problem is solved by spacing out several nuclei along the length of the fiber. Though muscle fibers are much longer than most other cells, they are seldom if ever more than a few centimeters long. To create the long muscles we see pictured in anatomi-

Human muscle fibers seen through a microscope. The stripes are the sarcomeres and the oval spots are nuclei; the color is a stain used to enhance contrast. These fibers have diameters of about 30 micrometers (0.03 millimeter).

cal drawings, fibers are arranged in bundles called fascicles, which may be much longer than the individual fibers. Similarly, woollen threads are very much longer than the fibers from which they are spun. The muscles at the back of an adult human's thigh have fascicles that can be as long as 250 millimeters, but it seems unlikely that any individual fiber is more than a small fraction of that length.

The fascicles in a muscle are able to develop tension and shorten, pulling on the bones and so moving the joints. Muscles can pull but they cannot push, so at least two muscles are needed to work a joint. For example, we bend our knees by shortening the hamstring muscles at the back of the thigh and straighten them by shortening the quadriceps muscles in front of the thigh. Shortening either of these groups of muscles stretches the other.

Muscles obviously cannot pull on a joint unless they are somehow connected to it. Some muscle fascicles connect directly to bone at their ends, but in many cases the attachment is made through tendons. In mammalian muscle (for example, beef) the muscle fascicles are red or pink and the tendons are white. The muscle fibers are the motors that power the body, and the tendons are simply the ropelike connections between the fibers and the bones.

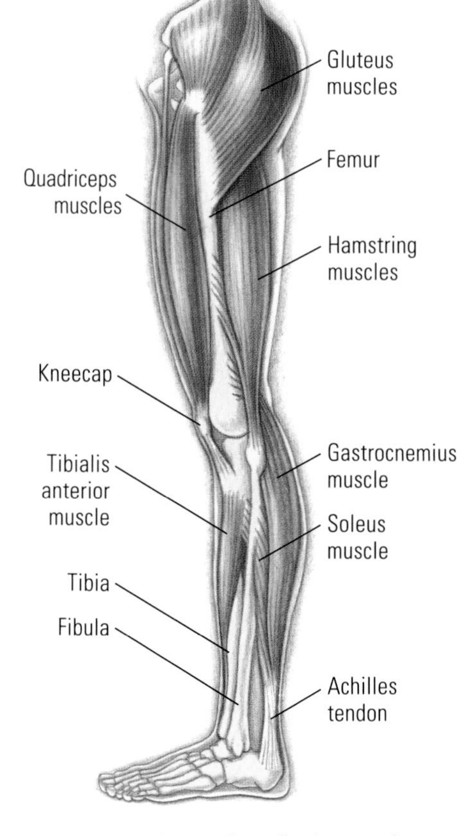


The principal muscles of a human leg.

Muscle fascicles run straight along almost the whole length of the hamstring muscles at the back of the thigh; only a little tendon is attached at each end. In contrast, the fascicles in some other muscles run obliquely between two or more long tendons. This "pennate" arrangement increases the force that the muscle can exert by packing in a larger number of fascicles. For example, the pennate muscles of the human calf (the gastrocnemius and soleus muscles), which connect by the Achilles tendon to the heel, can probably exert more than twice the force of the hamstrings even though they have only 0.9 times the volume. The inevitable penalty for packing in the extra fascicles is that they have to be short and so cannot lengthen and shorten very much. This restricts the range of angles through which the muscles can move their joints. As a rough general rule, muscle fascicles work well over a range of lengths in which the minimum is 70 percent of the maximum. A hamstring muscle fascicle, for example, might extend to a maximum length of 140 millimeters and contract to a minimum of 100, a range of 40 millimeters. A much shorter fascicle from the soleus muscle might extend only to 35 millimeters and contract to 25, a range of 10 millimeters. These ranges are not

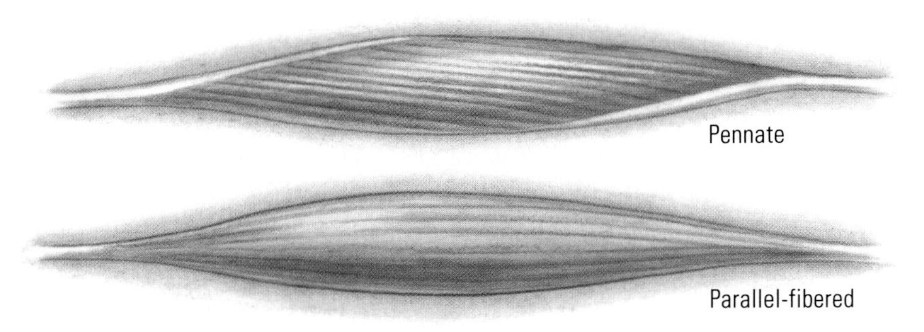


Muscle fascicles run the whole length of parallel-fibered muscles, but run obliquely between tendons in pennate muscles.

the maximum possible, but at longer and shorter lengths the forces that the muscles could exert would be severely reduced. The penalty for having a strong soleus muscle is that its short fascicles are effective only over a small range of length, allowing only a small range of ankle angles. A change in length of 10 millimeters in the soleus is only enough to move the ankle through an angle of about 12 degrees.

To understand how muscles shorten and why they can shorten only over a limited range of lengths, we need to look at the fine detail of their structure. We have seen that muscle fascicles are bundles of fibers, each of which is an individual cell. Within each fiber, occupying most of its volume, are an enormous number of fine filaments, much too fine to be seen by even the most powerful light microscope. Only electron microscopes can show the structure and arrangement of these filaments.

The filaments that occupy a fiber are of two kinds: extremely thin ones that consist mainly of the protein actin and thicker ones of the protein myosin. In the muscles that power the movements of vertebrates and most other animals, these filaments are arranged in a very regular repeating pattern: bands of thick and thin filaments alternate along the length of each fiber. Where the two kinds of bands overlap at their ends, thick filaments lie alongside thin ones. Midway along each band of thin filaments, partitions called Z discs cross the fiber. The repeating unit of pattern, the segment from one Z disc to the next, is called a sarcomere. Electron microscope pictures show that the filaments are the same length when the muscle is extended as when it is shortened; only the degree of overlap between thick and thin filaments changes. Think of the way the tubes of telescopes

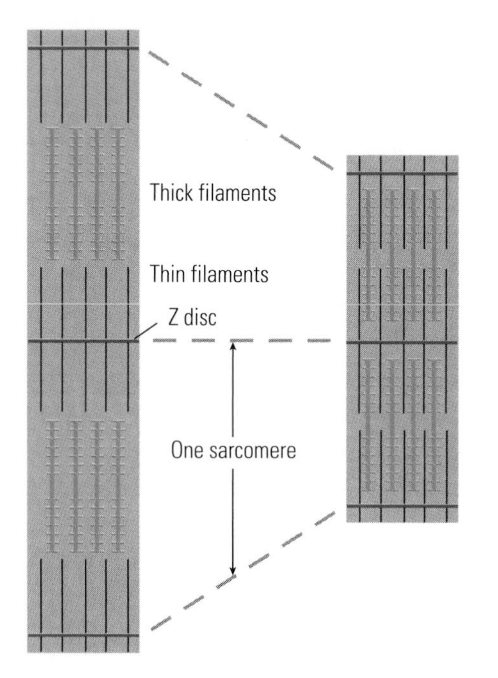


The thick and thin filaments in muscle fibers overlap less when the fiber is long *(left)* and more when it is short *(right)*. The muscle can exert maximum force when there is enough overlap for all the crossbridges to attach *(right)*. This diagram is based on electron microscope pictures.

overlap less when the telescope is extended and more when it is collapsed.

When high-resolution electron microscope pictures are examined closely, tiny arms called "crossbridges" can be seen projecting from the sides of the thick filaments. When the muscle is active, these crossbridges attach to neighboring thin filaments, pull, detach, and reattach, pulling the thin filaments along like a line of people pulling in a rope hand over hand. Obviously, they could not pull on the thin filaments if the muscle were extended so much that there were no overlap between thick and thin filaments—but muscles are protected from being stretched so much by a fine network of tendonlike material between the muscle fibers. When there is only a little overlap, only a few crossbridges can attach and the muscle can exert little force. When there is enough overlap for all the crossbridges to attach, the muscle can exert its highest force. Further shortening, however, results in less force, partly because crossbridges always pull toward the midpoint of the thick filament, and if the thin filaments overlap beyond this midpoint, some of the crossbridges pull the wrong way. Furthermore, if the muscle shortens too much, the thick filaments collide with the Z discs, impeding further shortening. That is why muscle fascicles work well only over a limited range of lengths.

Muscle fascicles are limited not only in the range of lengths at which they can work, but also in the rate at which they can shorten. The faster they shorten, the less force they can exert, and they can shorten at their highest speed only when there is no force resisting the movement. The reason is that crossbridges take time to detach, reposition themselves, and attach again. When a muscle is shortening rapidly, the crossbridges have to detach and reattach more frequently and, because this takes time, a smaller proportion of them are attached and pulling at any instant. Furthermore, if a crossbridge remains attached while the filaments slide past each other, it will eventually get into a position in which it is pulling the wrong way, resisting the shortening instead of helping it. The faster a muscle is shortening, the more crossbridges remain attached so long that they pull the wrong way.

The maximum shortening speed is usually stated in lengths per second. For example, a muscle that shortened by a quarter of its length in a quarter of a second would be described as shortening at a rate of one length per second. Different muscles have very different maximum shortening speeds: some muscles in the legs of mice can shorten at 20 lengths per second, whereas others in the legs of tortoises can probably shorten no more than 0.1 length per second. Muscles obviously must be fast enough to do the jobs required of them, but if muscles are unnecessarily fast energy is wasted, as we will see. If the muscles in the legs of mice were not able to shorten so fast, the mice could not run so fast and would be in more danger of capture from predators. Tortoises have no need to run to escape because they can retire into the safety of their shells, so very slow muscles that are very economical of energy are best for them.

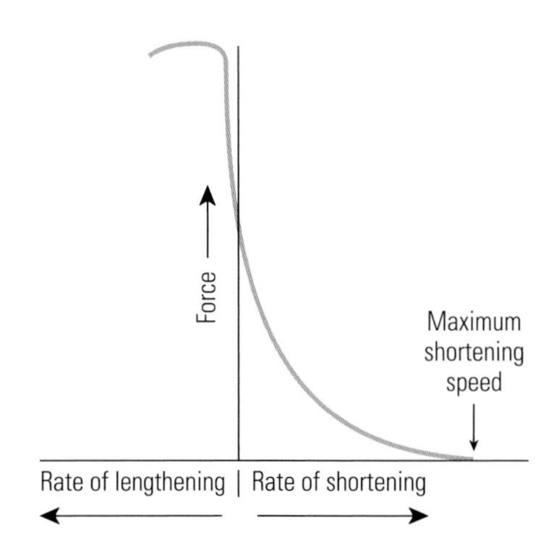


The faster a muscle shortens, the less force it can exert. This graph shows the results of experiments on frog leg muscles.

Like all bodily processes, muscular movement is powered by the energy that comes from food. For many animals, food is hard to obtain, and there is a strong advantage in using muscles and patterns of movement that are as economical of energy as possible. For this reason, we will want to know about the energy costs of different styles of locomotion and how energy is used by the muscles that power them.

## The Energy Costs of Movement

Muscles are fueled by the carbohydrates and fats obtained from food. Usually, these foodstuffs are oxidized by chemical reactions equivalent to burning:

$$\text{carbohydrate or fat} + \text{oxygen} \longrightarrow \text{carbon dioxide} + \text{water}$$

Such reactions release energy, some of which can be used to power muscular movement and the other processes of life. The amount of energy released can be measured in units called joules: for every liter of oxygen that reacts with carbohydrate or fat, about 20,000 joules of energy is released, from almost any foodstuff.

The link between oxygen use and energy is the basis of a very convenient method for finding out how fast animals use energy: if you measure their oxygen consumption, you can calculate how fast energy is being used. We will see later in this book how the method has been used to discover the energy costs of running, swimming, and even flight for a wide variety of animals. However, it does not always work because there are processes (as we shall soon see) that make food energy available to power muscles without any immediate use of oxygen.

The oxygen consumption method has also been used to find out how much energy different muscles need to do different things. Norm Heglund of Harvard University and Giovanni Cavagna of the University of Milan removed muscles from the dead bodies of frogs, toads, and rats and stimulated them electrically to activate them—that is, to set their crossbridges working, exerting force. They stretched the muscles or allowed them to shorten at controlled rates and measured how fast they removed dissolved oxygen from the surrounding solution. The muscles used oxygen at a low rate even when they were not stimulated and were exerting no force: some energy is needed to sustain life, even in inactivity. Whenever the muscles were stimulated and developed force, they used oxygen much faster. If they were held at constant length while being stimulated, they used oxygen many times faster than when at rest. If they were allowed to shorten, they exerted less force but nevertheless used oxygen as fast as or faster than when at constant length. These results tell us that

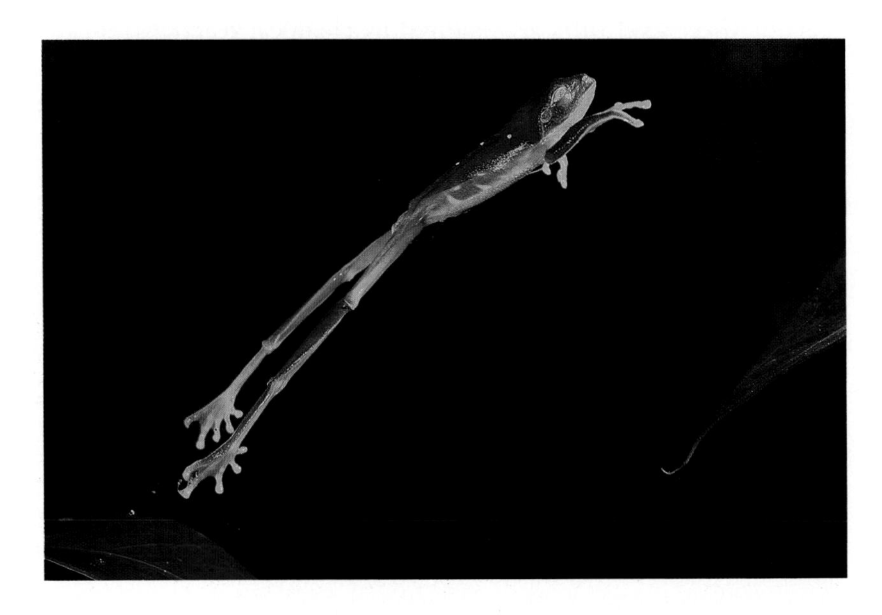

A red-eyed tree frog (*Agalythnis callidryas*) jumping between branches. The adhesive pads on its toes enable it to hold onto the leaves.

the faster a muscle is shortening, the faster it must use energy to maintain a particular force. Experiments by other people have shown that if a muscle is being stretched it can exert the same force for less energy cost. Why does a shortening muscle use energy faster than a lengthening or still muscle? Muscles can pull but they cannot push, so to power an animal's movement they must shorten. Only shortening muscles are doing work.

We need to be clear about the technical meaning of the word "work." Work is energy delivered by a force that moves the object to which it is attached: no work is done unless there is movement, and this movement must be in the direction of force. When I stand holding a suitcase, I am doing no work even though (if the suitcase is heavy) I may have to exert a large upward force to support it and my muscles may be using food energy quite fast. When I carry the suitcase across a level floor, I am still doing no work because my horizontal movement is at right angles to the vertical force: there is no movement in the direction of the force. If, however, I lift the suitcase higher off the floor, I am moving it in the direction of the force and doing work on it. Similarly, if I carry it up a hill, there is a vertical component to the movement and work is done. The work done by a force equals the force multiplied by the distance moved in the direction of the force:

$$\text{work} = \text{force} \times \text{distance}$$

Work is a form of energy, so it is measured in joules, the standard unit for *all* forms of energy. When a force of 1 newton moves an object a distance of 1 meter (in its own direction), it does 1 joule of work.

The notion that you are not doing work when you are straining to carry a heavy suitcase may seem odd. But we must realize that the scientific definition of work includes a particular relationship with energy. Whenever work is done, energy is given to the moved object or to its surroundings. When I stand holding a suitcase, no work is done and the suitcase's energy remains unchanged, just as if it were supported by an inanimate structure such as a table. When I carry the case across a level floor, its energy is again constant (provided I carry it at constant speed) and again (if we ignore air resistance) no work is needed. To understand why not, imagine a suitcase on a frictionless trolley on a level floor. If I gave the trolley a push to start it moving, it would continue to move at constant speed, with no need for any additional input of energy, until it hit an obstacle. Real trolleys slow down only because their energy is gradually dissipated by friction in their moving parts and by air resistance.

We still do not fully understand the relationship between the rate of energy consumption of a muscle and the job it is doing, but it seems helpful as a rough guide to think of the energy used by a muscle as the sum of two energy costs:

$$\text{total energy cost} = \text{cost of force} + \text{cost of work}$$

The equation shows why muscles use food energy faster when shortening than when exerting the same force and holding constant length. The extra energy is the cost of doing the work. The Principle of Conservation of Energy, one of the firmly established principles of physics, says that energy cannot be created from nothing (nor can it be destroyed). That tells us that for every joule of work that muscles do, they must use at least a joule of food energy. They actually use much more, commonly about 5 joules, because a lot of the food energy is lost as heat. (When we shiver we are using muscular movement to produce heat, to warm ourselves up.) The ratio of work output to food energy input is called efficiency, so if 5 joules of food energy is needed to do 1 joule of work, the efficiency is 1/5, or 0.2. Obviously, high efficiency is desirable, since it will lower the cost of work—the amount of food energy that has to be used in order to do the work:

$$\text{cost of work} = \frac{\text{work}}{\text{efficiency}}$$

The efficiency of a muscle is not constant but is different at different rates of shortening. As a general rule, the least metabolic energy will be used while doing a given amount of work if the muscle shortens at about one third of its maximum shortening speed.

Muscles use food energy when they are exerting force, even if they are holding constant length and so doing no work. Food energy is needed to support the activity of the crossbridges, which do not simply hold tight but continually detach and reattach. Crossbridges go through their cycles of detachment and reattachment whenever a muscle applies force. It is in this activity of the crossbridges that we seek the factors that must affect the energy cost of exerting force, while holding constant length. These factors determine the second energy cost that, together with the cost of work, contributes to the total energy cost.

The larger the force, the greater the cost, because more fascicles have to be activated. Also, the longer the fascicles, the greater the energy cost: it seems obvious that it must cost twice as much energy to exert a force in a fascicle 10,000 sarcomeres long as to exert the same force in a fascicle only 5000 sarcomeres long. Finally, it costs twice as much energy to maintain a force for 2 seconds as to hold it for only 1.

These factors (force, fascicle length, and time) are not the only ones that affect the cost of force. There is also another, the "economy" of the particular muscle:

$$\text{cost of force} = \frac{\text{force} \times \text{fascicle length} \times \text{time}}{\text{economy}}$$

The economy is different for different muscles. It is larger for muscles that can shorten slowly (such as tortoise leg muscles) than for ones that can shorten faster (such as mouse leg muscles). As a rough general rule, doubling the maximum shortening speed halves the economy. The reason is that the crossbridges of faster muscles detach and reattach more rapidly, even when the muscle is not shortening.

Muscles cannot do work without exerting force, so the distinction that I have made between the cost of work and the cost of force is rather artificial. However, I think it is a useful step toward understanding how muscles use energy. Unfortunately, our understanding of this fundamental aspect of muscle physiology is still imperfect.

## Aerobic and Anaerobic Muscles

Some muscles can only work aerobically, in the presence of oxygen. While aerobic muscles are active, oxygen must be supplied fast enough to oxidize the foodstuffs that they are using as fuel. Consequently, the rate of energy use is limited by the rate at which the

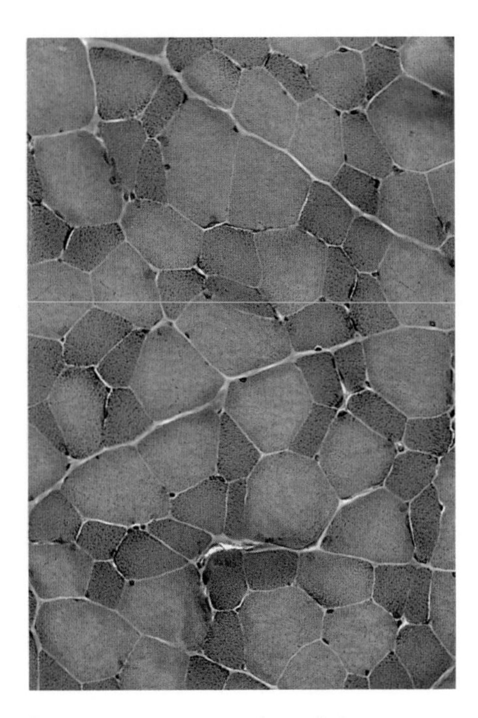

A microscope section through human muscle, stained to distinguish between aerobic fibers (darker stain) and the generally larger anaerobic fibers (paler).

lungs or gills can take up oxygen or the bloodstream can transport it. If an animal needs to make a burst of speed to, say, escape a predator or capture prey, oxygen may not reach its muscles fast enough. Fortunately, many animals have other muscles they can call upon, which work without oxygen.

Anaerobic muscles get energy from foodstuffs by processes that do not require oxygen—for example, by converting glucose to lactic acid. Anaerobic processes yield much less energy than does oxidation of the same quantity of food, so these processes would be very wasteful if the products were simply discarded. The more economical alternative is to oxidize some of the products afterward and use the energy so obtained to convert the rest back to carbohydrate. Thus an oxygen debt that has built up rapidly during a burst of violent activity may be repaid slowly while the animal recovers afterward. The advantage is that energy can be used faster to power a burst of activity. The disadvantage is that anaerobic processes cannot continue for long. When our muscles and those of other vertebrate animals work anaerobically, lactic acid accumulates, and we cannot tolerate more than a certain concentration of lactic acid in our bodies. Anaerobic processes enable us to make more intense bursts of activity than would otherwise be possible, but prolonged activities depend on aerobic muscle activity. For that reason, we cannot run marathons as fast as we can sprint. An athlete sprinting 100 meters uses mainly anaerobic metabolism, but one running a 42-kilometer (26-mile) marathon depends almost entirely on aerobic metabolism.

Some muscle fibers are specialized for aerobic work and some for anaerobic. The two types differ in biochemical composition, in microscopic structure, and often in color. In vertebrate animals, aerobic muscle fibers are generally reddish in color and anaerobic ones are generally whiter, but in cephalopod molluscs like octopuses and squids the colors are yellowish and white.

The color differences can be seen easily in some familiar foods. We are accustomed to seeing the swimming muscles of fish at our dinner tables in the form of fillets; these muscles consist mainly of white (anaerobic) muscle but generally have a band of red (aerobic) muscle running along the side, close under the skin. The red muscle is used for sustained swimming and the white for short bursts of speed, as will be explained in Chapter 7. The white breasts of chickens are the muscles that serve to beat the wings. Appropriately, they are anaerobic, since chickens make only very short flights. The darker leg muscles are largely aerobic to enable chickens to run around continually, using their leg muscles. Birds such as pigeons that make prolonged flights have dark muscles on their breasts, composed largely of aerobic fibers.

There are other distinct kinds of muscle with special properties that are important for their functions in locomotion—for example, there are the fibrillar flight muscles that enable small insects to beat their wings at extraordinarily high frequencies, which will be described in Chapter 5.

Many—perhaps most—animals use more energy for locomotion than for any other purpose. Locomotion is generally necessary for obtaining food, finding mates, and escaping predators. Because locomotion is such a large part of the energy budget, there is a big advantage to the animals in keeping energy costs low. So important are these costs that they will be considered repeatedly in the chapters that follow. Measurements of oxygen consumption will enable us to compare the costs of different forms of locomotion. The information in this chapter about how muscles use energy will help us to understand these costs. And we will see many remarkable ways in which evolution has adapted the structures of animals and their patterns of movement to make the costs of locomotion as low as possible.

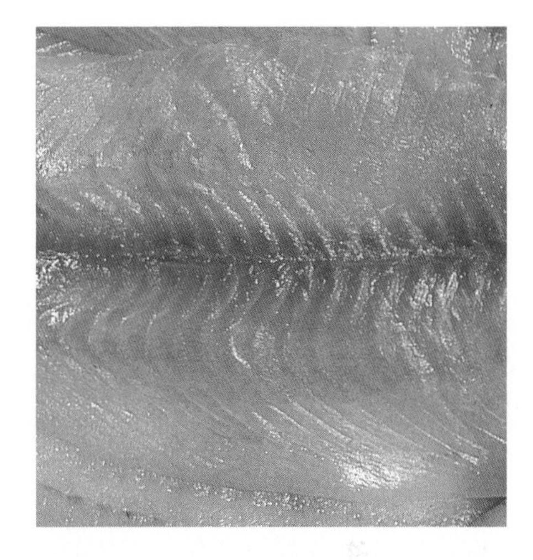

The darker stripe halfway down this fish fillet is the red muscle that powers sustained swimming. The remaining, paler muscle (the white muscle) is used only for short bursts of speed.

10
RUHLMANN

# 2

# Walking and Running

Successful racehorses gallop at 35 to 40 miles per hour. In this gait the two fore feet are set down in rapid succession, then the two hind feet, followed by the two fore feet again. The spring action of a sheet of tendon allows the back to bend and extend at appropriate stages, lengthening the stride.

Human walking is unique. No animal walks as we do; although birds walk on two legs and apes sometimes do, their styles of walking are quite different from ours. Most birds use flight rather than walking as their principal means of travel, and apes do most of their walking on four legs. Apart from ostriches and other flightless birds, no other animal depends as much as we do on two-legged walking. Is there something particularly good about our peculiar walking style?

## Walking: A Pendulum of Swinging Legs

The distinctive feature of human walking is that we keep each leg almost straight while its foot is on the ground. The sequence of photographs at the top of the page follows the course of movement through a single step. At stage (c) of a step, the supporting leg is straight and vertical, so the body is high. At stages (a) and (e), the legs are straight but sloping, so the body is lower. As a consequence of our legs' changing slopes, our heads bob up and down by about 40 millimeters (1.6 inches) in the course of each step. Although this bobbing may seem like a trivial side effect, it actually may reduce the energy cost of walking. A comparison of speed and height over the course of a step shows why.

As we walk, our feet exert forces on the ground. The illustration shows the directions of the forces, which have been recorded by means of force plates, instrumented panels set into the floor that give

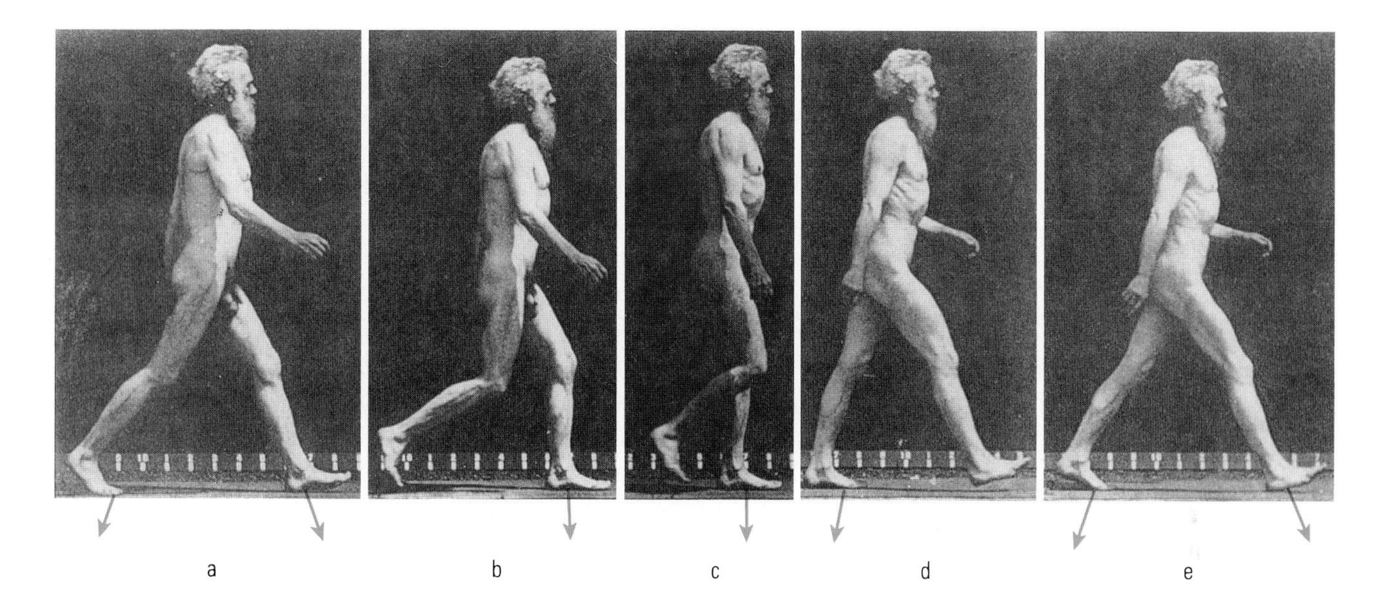


Stages of a walking stride, photographed in Eadweard Muybridge's studio in the 1880s. The pictures (showing, in this case, Muybridge himself) were obtained by firing still cameras in rapid succession. Arrows show the directions of the forces on the feet, as determined by more recent research.

electrical outputs indicating the downward, backward or forward, and sideways components of any force that acts on them. The records show that the force on each foot is always more or less in line with the leg. At stage (b) the foot is pushing forward as well as downward on the ground, so the body is not only being supported but is also being slowed down. At stage (d) the foot is pushing backward as well as down, so the body is being speeded up. Thus the body is traveling relatively fast at stages (a) and (e) of the stride and more slowly at stage (c). For example, the speed of someone walking moderately fast might fluctuate between 1.7 meters per second (3.8 miles per hour) at stages (a) and (e) and 1.4 meters per second (3.1 miles per hour) at stage (c).

To understand the implications of these movements and forces, we need to know about two kinds of energy: potential energy and kinetic energy. Potential energy is the energy that matter has because of its height. That energy changes with height is illustrated, for example, by hydroelectric schemes: as water flows downhill it loses potential energy, which is converted to electrical power. Kinetic energy is the energy that moving objects or fluids have because of their speed. For example, wind loses kinetic energy as it slows while passing over the blades of a windmill, and that energy supplies the power that drives the mill.

Whenever the body is raised it gains potential energy. That energy must come from somewhere; when you climb a mountain, for example, the potential energy you gain is supplied as work done by your muscles. The body also gains energy whenever it speeds up, and this kinetic energy must come from somewhere as well. When you accelerate at the start of a sprint, your muscles do the work that increases your kinetic energy. You might conclude that the muscles must do quite a lot of work in the course of each stride and that they consequently use a lot of metabolic energy. In walking, however, the body is high while it is traveling slowly (stage c) and low while it is traveling fast (stages a and e), so its potential energy is high while its kinetic energy is low, and vice versa. The same is true of a pendulum, which is highest when it stops at the end of a swing and lowest when it is moving fastest through the bottom of the swing. As the pendulum swings it converts potential energy to kinetic energy and back again. Energy is swapped back and forth between the two forms, and the pendulum will continue swinging for a very long time without any fresh input of energy. Similarly, very little work is needed from our leg muscles as we walk.

The pendulumlike quality of walking is a consequence of the straightness of our legs. It probably saves metabolic energy, but possibly not very much. If the changes in kinetic and potential energy were less well balanced, our muscles would have to do more work at some stages of the stride to increase the total (kinetic plus potential) energy of the body; at other stages they would have to work like brakes, doing negative work to reduce the (kinetic plus potential) energy. While doing positive work they would use metabolic energy faster, as explained in Chapter 1, but while doing negative work they would use it more slowly. The two effects might fairly nearly cancel each other out, but our knowledge of physiology is not precise enough for us to be certain. The pendulumlike quality of human walking probably reduces the "cost of work" element of the food energy requirement, but possibly not by very much.

There is another, possibly more important consequence of our straight-legged style of walking: it enables us to support our weight without the need for large forces in our leg muscles, thereby reducing the "cost of force" element of the energy requirement. When you stand with your knees straight, the line of action of your weight passes close to the knee joints and little tension is needed in your muscles to prevent the knees from collapsing under the load. If you stand with your knees bent, however, the line of your weight is farther from them and your muscles must exert more force. Similarly for walking: the straighter your legs, the less force your muscles need exert. For a practical demonstration, try taking a walk with your knees bent. You will feel the extra tension in the quadriceps muscles (at the front of the thigh), and you will find your steps unusually tiring.

Although our straight-legged style of walking is so economical of energy, only humans use it. Even our closest relatives, the apes, walk with their legs bent. Chimpanzees usually walk on all fours, on the soles of their hind feet and the knuckles of their hands, but they sometimes walk on their two hind feet only, especially when carrying things in their hands. Even when walking bipedally they walk unlike us, with their knees bent and their back sloping. Gibbons usually travel through the treetops by swinging from their long arms, which are too long for quadrupedal walking. They sometimes walk along the upper surfaces of thick branches, moving on their hind legs alone, with their knees bent. In the wild they seldom or never descend to the ground, but in zoos they sometimes do—and again walk bipedally on bent legs.

When a chimpanzee *(Pan troglodytes)* walks on all fours, it rests its knuckles on the ground, not the palms of its hands.

## Running: Bent Legs for Speed

Humans also move with bent legs—when they run. At the stage of a running step when the force on the foot is largest (stage c in the illustration on the next page), the knee is rather bent and the muscles must exert large forces, at a cost in metabolic energy. This extra cost suggests that running will be expensive of energy. If running is so expensive, and our straight-legged style of walking is so economical of energy, why do we ever run? Why don't we just walk faster?

Despite the apparent extra cost, adult people of normal size change from walking to running whenever their speed exceeds a highly predictable rate, about 2 to 2.5 meters per second (4.5 to 5.6 miles per hour). To try to explain why, we will analyze a simple

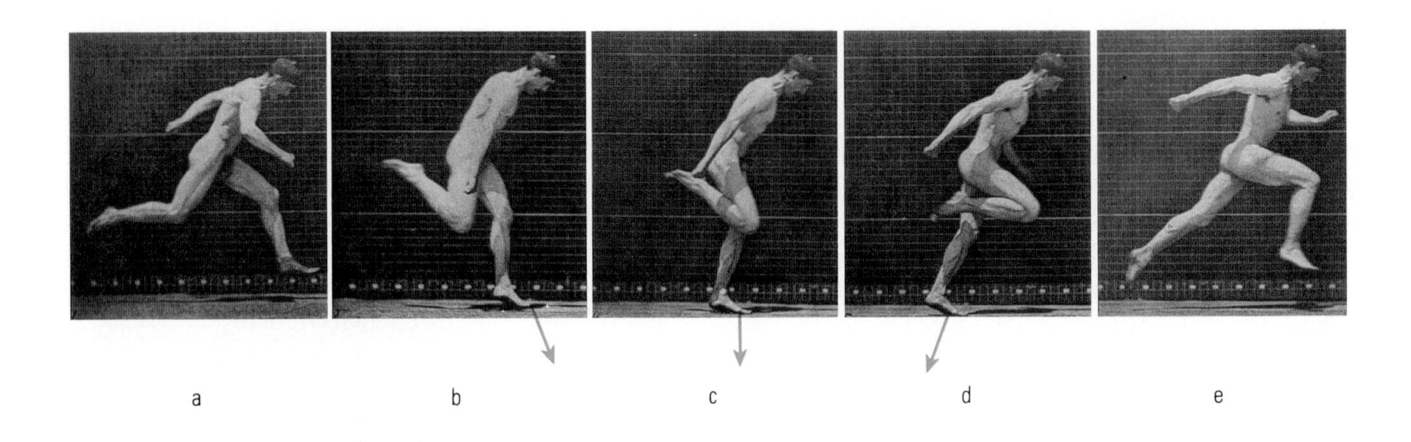


Arrows show the directions of the forces on the ground during the stages of a running stride, photographed by Eadweard Muybridge.

model that seems to reduce walking to its essentials. The person shown in the diagram on the facing page is walking at a speed $v$, keeping the legs utterly straight while the foot is on the ground. The body moves along an arc of a circle centered on the foot, a circle whose radius is the length $l$ of the leg. Here we need a simple formula from physics. An object moving in a circle is constantly changing its direction, turning toward the center of the circle. That change in direction affects the object's velocity, its speed in a specified direction. Because its direction is changing even though its speed may be constant, its velocity is also changing: in other words, the object has an acceleration toward the center of the circle. If the speed is $v$ and the radius $l$, this acceleration is $v^2/l$. A force pulling toward the center of the circle is needed to give the object this acceleration and so prevent it from flying off at a tangent. In the case of the walker, the acceleration is downward toward the foot and the force causing it must be the weight of the body, so the acceleration cannot be greater than the gravitational acceleration $g$:

$$v^2/l \text{ cannot be greater than } g,$$
$$\text{so } v \text{ cannot be greater than } \sqrt{gl}.$$

The gravitational acceleration $g$ is about 10 meters per second squared, and the length of an adult human leg from hip to sole is about 0.9 meter. Thus $\sqrt{gl}$ is about $\sqrt{10 \times 0.9} = 3$ meters per second. We can conclude that it is physically impossible to walk like the person in the diagram at speeds greater than 3 meters per second.

That is the maximum possible walking speed, and people actually change from walking to running at the slightly lower speed of about 2 to 2.5 meters per second. The maximum speed is lower for children because their leg length $l$ is shorter—and small children often have to run to keep up with their walking parents.

That seems to be a convincing explanation of the change from walking to running, until you look at walking races. The rules require the leg to be straight while the foot is on the ground, but athletes nevertheless manage speeds of 4 meters per second. The secret is a peculiar movement of the hips, which lowers the body's center of gravity a little at the stage of the stride when the leg is vertical. The body rises and falls less in each step than we supposed in the simple theory, so the accelerations that are needed at any particular speed are less than we calculated.

If you are not constrained by the rules of a walking race, it is best to change from walking to running at a speed a little less than the

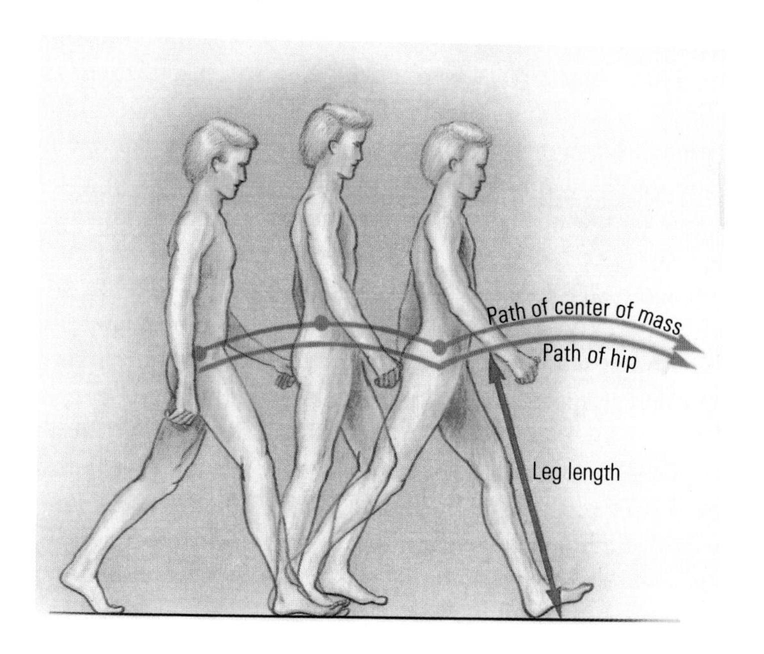


When a person walks on rigid legs, the body moves along an arc of a circle centered on the foot.

The peculiar hip movements of the racing walk enable athletes to walk faster than the maximum speed suggested by simple theory. The pelvis is tilted at the stage of the stride when the supporting leg is vertical (competitors 629 and 801).

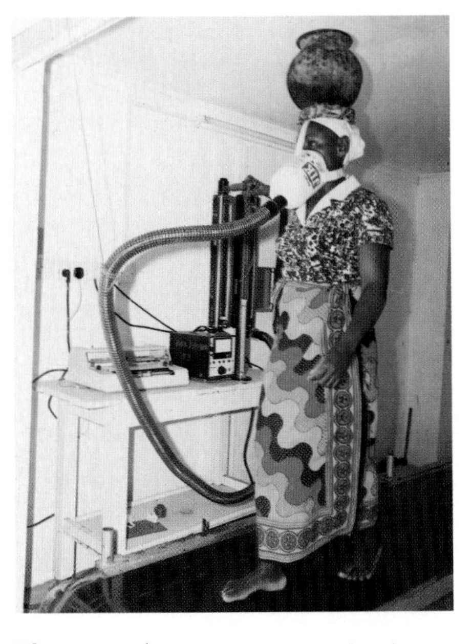

This woman's oxygen consumption is being measured while she walks on a moving belt.

theoretical maximum. The reason is that the energy cost of walking rises steeply at speeds near the limit. Physiologists have shown how the advantage shifts from one gait to the other by measuring the rates of oxygen consumption of people walking and running at different speeds. The subjects walk on a moving belt that keeps them stationary relative to the laboratory, while wearing face masks from which air is sucked through a hose to oxygen-analysis equipment. Plenty of fresh air for breathing is drawn in around the edges of the loose-fitting masks, but all the air the subjects breathe out is sucked away for analysis. The equipment measures the volume of air that passes through and the concentration of oxygen in it, so it is easy to calculate the volume of oxygen that has been removed by respiration, and from that the rate at which metabolic energy is being used. When energy consumption is plotted against speed for walking and then against speed for running, the resulting curves cross each other at about 2 meters per second. Below that speed, walking is the less energy consuming way of getting around; above it, running is more economical. At any particular speed we use the gait that needs less energy unless there is some special reason (such as the rules of a walking race) for doing otherwise.

Notice that both graphs give rates of energy consumption only for low to moderate speeds. The reason is that at higher speeds muscles work anaerobically and we build up an oxygen debt. Measuring oxygen consumption is a good way of finding out how much energy is needed for locomotion, but only at speeds at which the muscles are working aerobically.

## The Spring in a Running Step

Runners bounce along in a motion that looks quite different from walking. Running also depends on a different energy-saving principle. When we walk, each foot is on the ground for more than half the stride, so there are stages when both feet are on the ground simultaneously. In contrast, when we run, each foot is on the ground for less than half the stride, so there are stages when both feet are off the ground. Thus running is a series of leaps, and the body is highest when the feet are off the ground, at stages (a) and (e) of the illustration on page 22. As in walking, the forces on the feet keep more or less in line with the legs, so the body is slowing down at stage (b) and speeding up at stage (d). Thus the body is higher and moving faster at stages (a) and (e) but is lower and moving more slowly at stage (c). Its kinetic and potential energies are highest at stages (a) and (e) and lowest at stage (c). There can be no question here of energy being swapped back and forth between the two forms in the pendulumlike manner of walking, but we will soon see how energy is saved by a different energy-swapping principle.

The basic principles of human running are the same as those of kangaroo hopping, even though the two forms of motion look very different. The kangaroo sets both hind feet down on the ground at the same time, but during each hop kinetic and potential energy rise and fall together as in running: they are highest in midair and lowest at the midpoint of the period of contact of the feet with the ground. Again, there is no question of a pendulumlike exchange of kinetic and potential energy. The same is true of the running gaits of quadrupedal mammals (dogs, horses, antelopes, and all the rest), which will be described later in this chapter.

In all these gaits, the animal travels like a bouncing ball. When a ball hits the ground, it is brought rapidly to a halt, losing kinetic

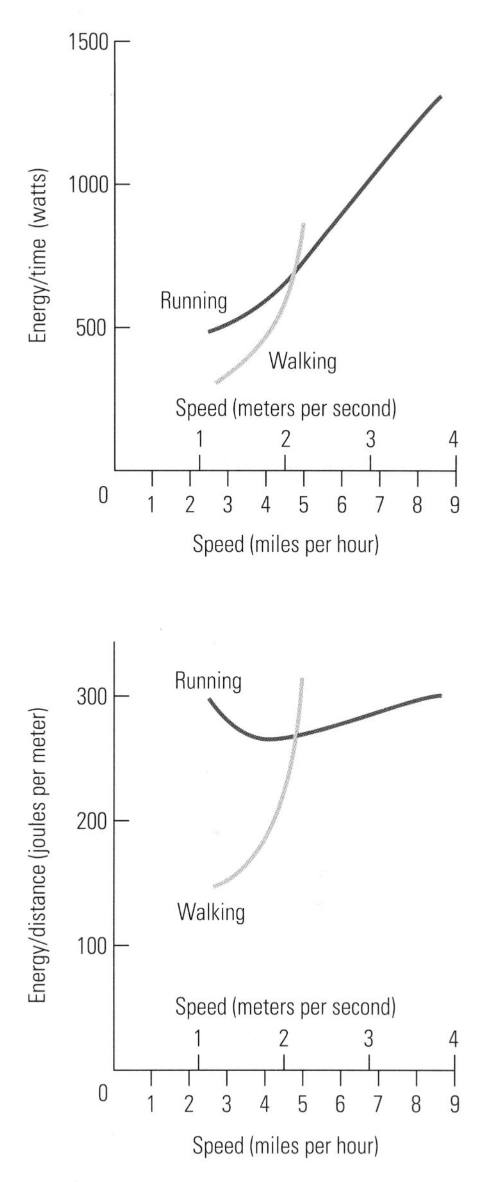


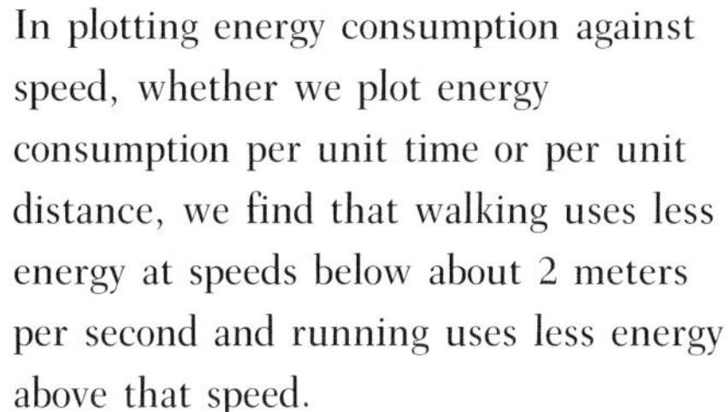

In plotting energy consumption against speed, whether we plot energy consumption per unit time or per unit distance, we find that walking uses less energy at speeds below about 2 meters per second and running uses less energy above that speed.

energy, which is largely converted into elastic strain energy as the ball is squashed out of shape. The ball then springs back to its original shape, and the elastic energy is converted back into kinetic energy as the elastic recoil throws the ball back into the air. Similarly, in the case of a running person or a hopping kangaroo, the kinetic energy and potential energy lost at each footfall are converted briefly into elastic strain energy and then returned in an elastic recoil.

Our bodies owe their bounce to springs that stretch to store elastic strain energy and recoil to return it. The most important of these springs are tendons, especially the tendons of muscles in the lower parts of the legs. As the connection between muscle and bone, tendons transmit the force from muscles to the moving joint. The force stretches the tendons whenever it increases and allows them to shorten whenever it falls. Tendons are not very obviously elastic: they stretch only a little before they break. In this respect they are more like ropes than like rubber bands. Yet the small amount of stretch is enough to save the muscles a considerable amount of work.

Tendons from different mammals and different parts of the body all have very similar properties, and all will stretch by about the same percentage of their length when under the same stress. Their elastic properties are best investigated in machines of the kind that engineers use to test the strength and elasticity of metals and plastics. At the top of the machine shown in the picture on this page there is a

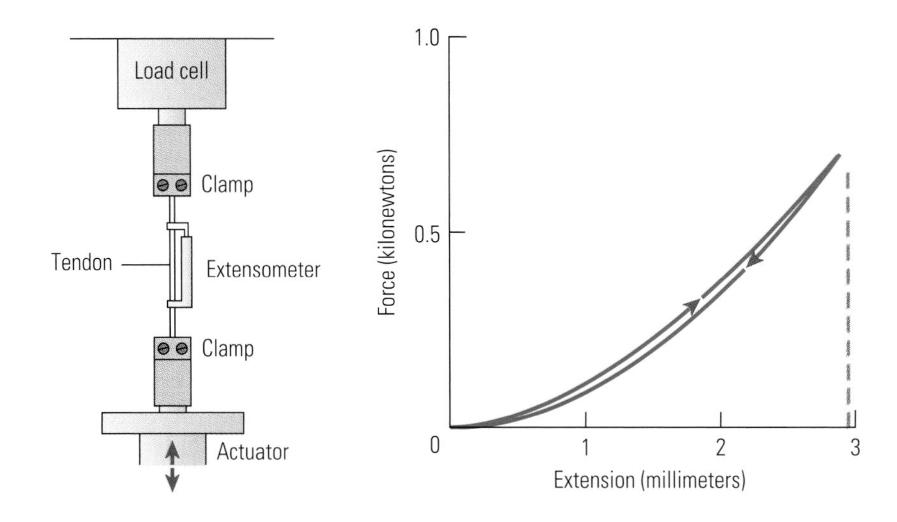


A dynamic testing machine used to measure the elastic properties of tendons *(left)*, and the record of a test on a tendon from the hind leg of a wallaby *(right)*.

load cell, an electrical device that measures any forces that are exerted on it. At the bottom is a hydraulic actuator that can be made to move up and down and that is capable of exerting large forces. A tendon dissected from an animal carcass is fixed in the machine, one end held in a clamp attached to the load cell and the other in a clamp attached to the actuator. Moving the actuator down stretches the tendon and moving it up allows the tendon to recoil. The machine can be adjusted so that the forces and the rates of stretching are about the same as they would be in running. The tendon would be moist inside the body and must be kept moist in the experiment because its elastic properties would change if it were allowed to dry out. The tendon may be kept at the temperature of the living body, but the extra bother generally seems unnecessary because the properties of tendons at normal room temperature are almost exactly the same as at body temperature.

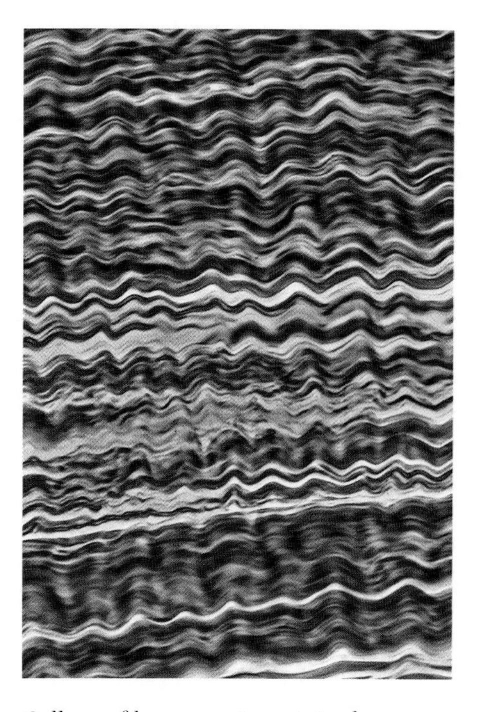

Collagen fibers seen in a stained microscope section of tendon. When a tendon is stretched, the wavy fibers straighten out. Each wave is about 0.3 millimeter long.

The graph shows the result of a test on a tendon from the hind leg of a wallaby, a small kangaroo. The record shows the force increasing as the tendon is stretched and falling as it is allowed to shorten, just as if a rubber band or a steel spring were being stretched and allowed to recoil. The line showing the force during stretching is slightly above the line for the recoil, so the record forms a narrow loop rather than a straight line. The loop in the record shows that the energy returned in the recoil is a little less than the work needed to stretch the tendon, which is inevitable because no material is perfectly elastic. (The energy that is not returned is lost as heat.) A tendon returns 93 percent of the energy, losing only 7 percent as heat; this percentage is good in comparison to rubbers and plastics. If it were a less good elastic material, more energy would be needed for running, because muscles would have to do work to replace the lost energy. Moreover, our tendons would heat up when we ran and might even get cooked.

The most important tendon for human running is the Achilles tendon, which you can feel through your skin, running behind your ankle to attach to the heel bone. To find out how much energy it can save, we need to know how much force acts on it and how much it stretches, because the strain energy stored in a spring is proportional to the force multiplied by the amount of stretch. It would be difficult to measure how much the tendon stretches when we run, but it is fairly easy to calculate that stretch.

The forces on a human foot at the stage of a running stride at which they are greatest.

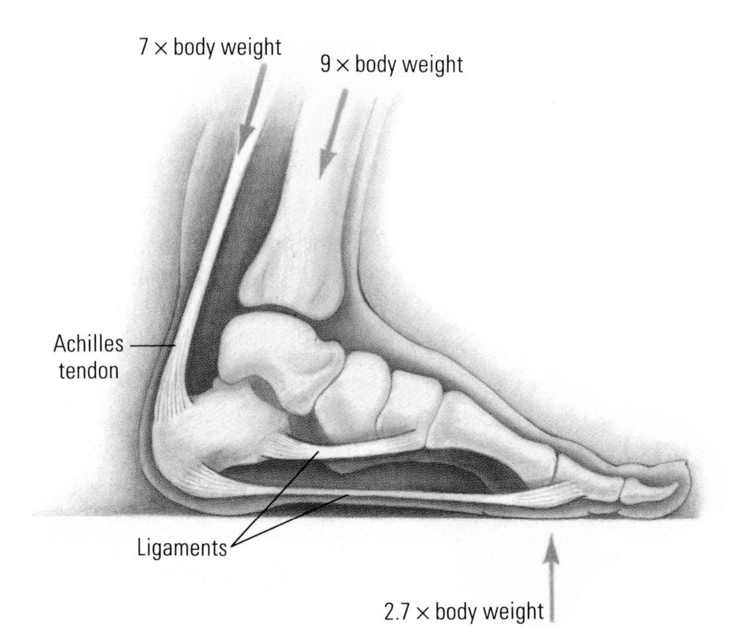


From records of people running across a force plate at middle-distance speeds, we find that each foot exerts a peak force on the ground of about 2.7 times body weight. The foot presses down with this force on the ground, and the ground pushes up on the foot with an equal, opposite force. This force is distributed over much of the sole of the foot, but for the purpose of calculation we can think of the entire force as acting at a single point, called the center of pressure. Sophisticated force plates tell us not only the size and direction of any force that acts on them, but also the position of the center of pressure. At the stage of a running stride that we are considering, the center of pressure is on the ball of the foot, close to the bases of the toes. The ankle joint is a freely movable pivot, so this force of 2.7 times body weight acting in front of it needs a balancing force behind it. (Similarly, the weight of a child on one side of a seesaw must be balanced by another child on the other side.) The balancing force on the foot is supplied by the muscles of the calf pulling upward on the heel bone through the Achilles tendon. The Achilles tendon is about 47 millimeters from the ankle joint, and the line of action of the ground force is about 116 millimeters from the ankle joint, so by the principle of levers the force in the tendon must be $(116/47) \times 2.7$, or almost 7 times body weight. This force is about 3500 newtons, or a third of a ton, for a typical (50-kilogram) woman and 5000 newtons, or half a ton, for a 70-kilogram man.

The cross-sectional area of the Achilles tendon in adult men is about 90 square millimeters, so the 5000-newton force sets up a stress in the tendon of about 56 newtons per square millimeter (8000 pounds force per square inch). This stress is about half the stress that would be needed to break the tendon, and it is enough to stretch it by about 6 percent of its length. If we include the part of the tendon that runs up into the flesh of the calf muscles, the tendon is about 250 millimeters long. When stretched by 6 percent, it extends about 15 millimeters, enough to allow the ankle to bend through 18 degrees. If the tendon were not extensible, the muscle fascicles would have to lengthen and shorten this much more to allow the ankle to make the movements that are needed for running. About one third of the negative and positive work that would otherwise have to be done by the muscle lengthening and shortening is done by the passive stretch and recoil of the tendons. If the tendon were inextensible, all the kinetic and potential energy lost by the body in the first half of the step would have to be removed by the muscles doing negative work and lost as heat. It would then have to be replaced by the muscles doing positive work in the second half of the step. Because the tendon is extensible, however, one third of the energy that would otherwise be lost is stored and returned.

Because the muscles do less work, metabolic energy is probably saved, but even further savings are possible. If the muscle fascicles need not lengthen and shorten so much, the person or animal can make do with muscle fascicles that are shorter. Alternatively, if the fascicles are long they need not shorten so fast, and the leg may be moved by slower, more economical muscle fibers. In either case, the "cost of force" element of the metabolic energy consumption will be reduced, as explained in Chapter 1.

In antelopes, horses, and related mammals, the tendons that serve as springs in running are very long and the muscle fascicles exceedingly short. Almost all the movement at the ankle joint, while the foot is on the ground in running, results from the stretching and recoil of the tendons, and the muscle fascicles lengthen and shorten very little. The most extreme example I know of is the plantaris muscle of the camel. The plantaris is rudimentary or even absent in

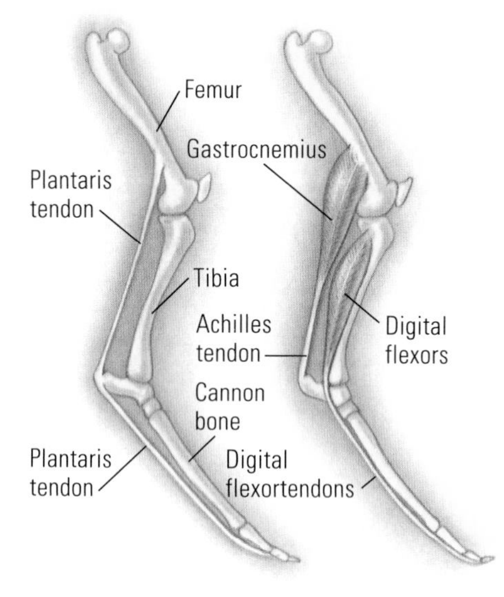


The hind leg skeleton of a camel, with some of the tendons that serve as springs, and their muscles (tinted pink).

people, but in most other mammals it is one of the strongest muscles of the hind leg. It runs from behind the knee, down the shank, around the heel, and along the foot to the toes. In the camel, its muscle fascicles have almost disappeared. Those that remain are only about 2 millimeters long, buried in the tendon, and they can surely have no significant function. The tendon itself is about 1.3 meters (51 inches) long, continuous from the knee to the toes, and like all tendons must serve as a passive spring. Here is a "muscle" that can exert forces and allow joints to move without any metabolic energy cost.

An arrangement so economical must have a drawback; otherwise all mammals would take advantage of it. In this case, the penalty for saving energy is loss of agility. If a camel's leg is positioned so that the plantaris tendon is taut, the camel cannot bend its ankle joint unless it also bends the knee or the toe joints to slacken the plantaris. This loss of freedom of movement is one of the many reasons that camels are not good at climbing trees.

The Achilles tendon is the most important spring in the human leg, but films of barefoot runners suggest that there is another spring in the foot. The films show that while pressed against the ground the foot is considerably squashed: the ankle is forced about 10 millimeters nearer the ground than if the foot were resting lightly. The flattening of the foot is a consequence of its arched structure. We have already seen that large upward forces act on the ball of the foot and (through the Achilles tendon) on the heel. These upward forces are balanced by a downward force of nine times body weight at the ankle joint, where the tibia (the principal bone of the lower leg) presses down on the joint. The two upward forces and the downward force in between partly flatten the arch of the foot, stretching some of the ligaments that connect the foot bones to one another.

My colleagues Robert Ker and Mike Bennett and I suspected that the arch of the foot might be a spring and wanted to test this idea experimentally. The machine that we had used for stretching tendons was also suitable for squeezing feet, but we could think of no way to use the machine safely on feet that were still in place on people's legs. Instead we used feet that had been amputated by surgeons because they were diseased or, in one case, because the knee had been damaged beyond repair in a traffic accident. The tibia was attached to the machine's load cell and the foot rested on two steel blocks, which in turn were supported by the actuator. When the actuator was made to rise, the foot was squeezed and the arch flattened. Rollers below the uppermost steel blocks allowed the slight lengthening of the foot that occurred when the arch flattened.

This simple experiment imitated quite well the forces that act on the foot during running. The upward pressure from the block under the ball of the foot imitated the force from the ground. The other block pressed directly on the heel bone (we had removed the fatty pad of the heel, to expose the bone), and its upward push imitated the upward pull of the Achilles tendon. Finally, the downward push from the load cell imitated the force at the ankle joint of the living foot. We made the actuator move up and down once or twice per second.

The experiments confirmed that the foot is indeed a reasonably good spring. It compressed under load and recoiled immediately when the load was reduced. The records showed loops that were wider than those in the experiments on tendon, indicating that more of the energy was being lost as heat, but most of the energy (78 percent) was returned.

We have already estimated that one third of the kinetic and potential energy that the body loses and regains in a running step is stored

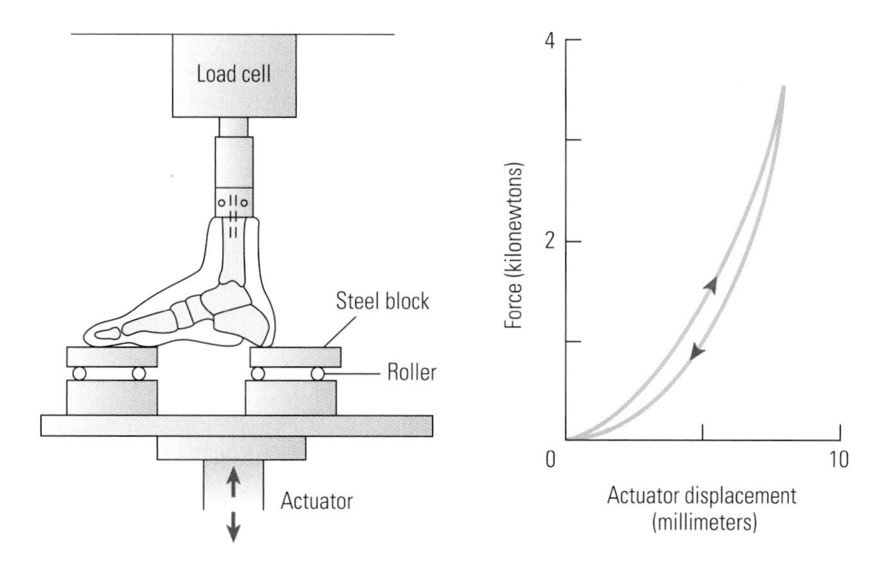


The experiment that demonstrated the spring in the arch of the human foot *(left)* and a typical result *(right)*.

in the Achilles tendon and returned in its elastic recoil. The experiments on feet showed that they would store and return a further one sixth of the kinetic plus potential energy. Together, these two springs halve the work that the muscles have to do ($\frac{1}{3} + \frac{1}{6} = \frac{1}{2}$). The tendon springs of animals such as horses, camels, and antelopes are probably even more effective.

## Four-Legged Gaits

Most animals that travel on two legs have a slow gait and a fast one. People and some birds (chickens, for example) walk to go slowly and run to go fast. Other birds, such as American robins, walk at low speeds, but at higher ones they hop. Kangaroos also have two gaits (the hop is the faster one), but in their case the slow gait is an awkward shuffle on all four legs and the tail. The walk, the run, and the hop exhaust the possibilities for two-legged gaits, but much more variety is possible when four legs are used. Most quadrupedal mammals of the size of cats or larger switch back and forth between three different gaits: the walk, the trot, and the gallop (although really small mammals such as mice seldom walk).

In the quadrupedal walk, as in the human one, each foot is on the ground for more than half the time. The four feet are set down in turn, generally at roughly equal intervals of time and almost always in this order: left fore, right hind, right fore, left hind, left fore, and so on. We will examine the possible significance of the sequence later in the chapter, when we consider stability. The legs are not kept as straight as in human walking, and the pendulum effect is less pronounced.

Trotting and galloping are running gaits: each foot is on the ground for less than half the stride. There may or may not be a stage in the stride when all four feet are off the ground, but at least there are stages when both fore feet, or both hind feet, are off. In trotting, the feet move in diagonally opposite pairs, the left fore with the right hind and the right fore with the left hind. Camels and some long-legged breeds of dog use the pace, a gait that is superficially trotlike but in which the two left feet move together and the two right feet move together, instead of the diagonal pattern. Perhaps these animals

Four stages of a stride of a walking horse *(top)*, a trotting horse *(middle)*, and a galloping horse *(bottom)*, photographed by Eadweard Muybridge. The walking horse moves its four feet in turn in the order left hind, left fore, right hind, right fore; the trotting horse moves diagonally opposite pairs of feet together; and the galloping horse sets down first one pair of feet, and then the other.

prefer the pace because if they trotted, the long legs might get in each other's way at the stage when the fore leg swings back and the hind leg of the same side swings forward.

In walking and trotting the left and right feet of a pair are set down at equal intervals; for example, if the right fore foot is set down a quarter of a second after the left, the left is set down a quarter of a second after the right. In a gallop, the intervals are unequal. The two feet of a pair are set down in rapid succession, and then a longer interval follows before the first is set down again. In a full gallop, the two hind feet are set down and then the two fore. The canter, which is sometimes recognized as a distinct gait, is a slow gallop in which the first fore foot is set down at the same time as the second hind.

The footfall pattern of galloping makes it possible for the animals to lengthen the stride by bending and extending the back. While only the fore feet are on the ground, the back bends, pulling the hindquar-

Four stages of a stride of a Bactrian camel *(Camelus ferus)*. This gait is called the pace: legs on the same side of the body move together.

ters forward. While only the hind feet are on the ground, the back straightens again, pushing the forequarters forward. Thus the body moves farther forward while each pair of feet is on the ground than it would if the back remained rigid. That enables the animal to travel faster or (as we will see) to travel more economically at the same speed.

The principal muscle that straightens the back is connected to the skeleton of the hindquarters by a sheet of tendon (the technical term is *aponeurosis*). This aponeurosis has elastic properties like other tendons and serves as an energy-saving spring, but only in galloping. To see why a spring might be useful, we have to think more about the animal's kinetic energy. We have seen how the body as a whole decelerates and reaccelerates while the feet are on the ground, but we have ignored the movements of the legs, which swing back while their feet are on the ground and forward again while they are off. Kinetic energy is associated with this movement whenever the legs are moving *relative to the body's center of gravity*. At the end of its forward swing and again at the end of its backward one, each leg has to stop and start swinging the opposite way: it has to lose and regain kinetic energy twice in each stride. The faster an animal runs, the faster the legs have to swing and the larger the swings in the kinetic energy. At the stage of the stride when the back is most bent, the fore legs have been swinging back and are about to swing forward and the hind legs have been swinging forward and are about to swing back. Both pairs of legs have to be stopped and started moving again, so kinetic energy has to be lost and regained. That could be accomplished entirely by muscles doing negative work to stop the legs and then positive work to reaccelerate them, but less metabolic energy is needed if some of the kinetic energy is stored as elastic strain energy in the aponeurosis and returned by its elastic recoil. That is what seems to happen.

We have already seen how, for people, walking is more economical than running at speeds below 2 meters per second, whereas running is more economical at higher speeds. Dan Hoyt and Richard Taylor of Harvard University showed similarly for ponies that each of the three gaits—walk, trot, and gallop—was the most economical in the range of speeds at which it is used. Just as in the experiments with people, they had the ponies run on a moving belt while the air they breathed out was collected through face masks and analyzed. It is fairly easy to train ponies (and many other animals) to run on a moving belt, but Hoyt and Taylor achieved the more difficult feat of training the ponies to walk, trot, or gallop on command. These ponies could be made to gallop at speeds at which they normally would have trotted and to trot at speeds at which they would have preferred to gallop. By analyzing the use of oxygen Hoyt and Taylor obtained a graph of energy consumption per unit distance plotted against speed. The walking and trotting curves cross at 1.7 meters per second, telling us that walking is more economical below that speed and trotting above. Similarly, the trotting and galloping curves cross at 4.6 meters per second; above that speed the advantage shifts to galloping. To travel as economically as possible, the ponies should have changed from walking to trotting at 1.7 meters per second and from trotting to galloping at 4.6 meters per second.

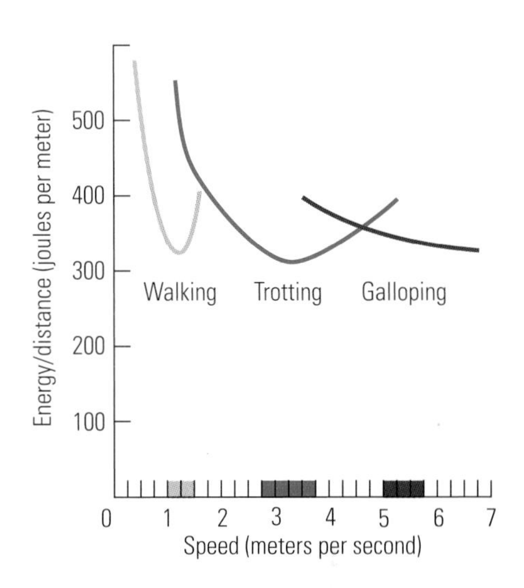


The energy used by ponies as they walked, trotted, or galloped at various speeds. The boxes at the bottom of the graph show the range of speeds at which the ponies used each gait when moving freely in their field.

To find out whether the ponies did that, Hoyt and Taylor filmed them moving around their paddock. The two men did not chase the ponies or disturb them in any way, but simply allowed them to move as they chose. They found that the ponies did indeed select the most economical gait, walking below 1.5 meters per second, trotting at 2.8 to 3.8 meters per second, and galloping above 5 meters per second. Furthermore, the ponies generally avoided speeds near the intersections in the graph: they accelerated quickly from walking well below 1.7 meters per second to trotting well above it, and from trotting well below 4.6 meters per second to galloping well above it.

The reason is that ponies need to use less energy per unit distance near the middle of the speed range for each gait and must use more near the transition speeds. For example, a pony that travels at 4.6

meters per second would use 340 joules per meter if it moved steadily at that speed (whether trotting or galloping), but it could cover the same distance in the same time for less energy if it alternated between trotting at 3.5 meters per second and galloping at 6 meters per second using (in each case) only about 300 joules per meter.

People similarly avoid speeds near the walk-run transition. Ultrarunning is the sport of racing over a distance of 100 miles. The best competitors cover the distance in about 13 hours, at a speed of about 3.4 meters per second, running all the way, but many others average speeds of around 2.2 meters per second. These latter competitors walk part of the way at lower speeds and run the rest at higher speeds, avoiding the uneconomical transition speed.

Not surprisingly, different animals change gaits at different speeds: for example, short-legged cats make the changes at lower speeds than do long-legged giraffes. To discover the rule that relates gait-change speeds to size, we have to look more generally at the consequences of size differences for four-legged animals.

## Walking, Running, and the Design of Ships

Imagine that small animals were exact scale models of large ones. To say the same thing in more technical terms, imagine that animals of different sizes were geometrically similar to one another. If one animal were twice as long as another, it would also be twice as wide and twice as high, and all its bones would be twice as long and have twice the diameter. That is obviously not the case: a 2-kilogram cat is not an exact scale model of a 250-kilogram tiger, nor is a 20-kilogram gazelle an exact scale model of an 800-kilogram buffalo. However, it is more nearly the case than you might suppose, as a simple argument will show. The masses of geometrically similar animals would be proportional to the cubes of their lengths (our imaginary animal that was twice as long, twice as wide, and twice as high as another would have $2 \times 2 \times 2 = 8$ times the volume and so be 8 times as heavy). In other words, the lengths of geometrically similar animals would be proportional to the cube roots of their masses, or to $(\text{body mass})^{0.33}$. With colleagues in Britain and Kenya I measured the leg bones of

The galloping movements of cats and rhinoceroses are remarkably similar even though the animals are so different. These outlines have been traced from films.

mammals ranging from 3-gram shrews to a 3-tonne elephant. (The metric tonne of one million grams is only a little different from the ton used in commerce.) We found as a general rule that the lengths and diameters of the leg bones were approximately proportional to (body mass)$^{0.36}$, which is little different from the rule for geometric similarity.

From similarity of shape we move to similarity of movement, for by extending the idea of geometric similarity we can arrive at the idea of dynamic similarity. Two shapes are geometrically similar if they could be made identical by uniform changes in the scale of length. Two motions are dynamically similar if they could be made identical by uniform changes of the scales of length, time, and force.

To understand what that means, imagine you had a film of a small cat running and another of a tiger running. You could change the sizes of the images on a screen by moving the projectors closer or farther away. You could also change the time taken for each stride by running the projectors faster or slower. If the animals were running in dynamically similar fashion, you would be able to make the two films seem identical.

The movements of cats are remarkably like those of tigers using the same gait. Indeed, there are close similarities of movement even between less similar animals; for example, between cats and rhinoceroses. Would it be reasonable to suggest as a rough approximation that different mammals using the same gait move in dynamically similar fashion?

There is a physical principle that says that movements that are affected by gravity cannot be dynamically similar unless their Froude numbers are equal:

$$\text{Froude number} = \frac{(\text{speed})^2}{\text{gravitational acceleration} \times \text{length}}$$

The principle was first used by William Froude, a Victorian naval engineer who was making tests on small-scale models before building ships to new designs. He wanted to know how much power would be needed to propel the full-sized ships. Much of this power is needed because ships push waves along in front of their bows. To find out what the waves would be like around the real ships, Froude needed to know how fast he should propel the models to produce the same pattern of waves. He showed that the model speed should be adjusted to make the Froude number the same as for the ship. The wave patterns around model and ship would then be dynamically similar. As always when the Froude number is applicable, gravity is important here—in this case because it tends to flatten the waves. The same principle has since been applied to other situations in which gravity is important, and I was the first to apply it to walking and running.

The speed and the length that are used to calculate the Froude number can be defined in various ways to suit different kinds of movement. When considering walking and running, it seems sensible to use the forward speed of the body and the length of the legs:

$$\text{Froude number} = \frac{(\text{speed of locomotion})^2}{\text{gravitational acceleration} \times \text{leg length}}$$

We can expect different animals to run in dynamically similar fashion only at speeds that make their Froude numbers equal. For example, camels' legs are about nine times as long as those of cats, so the Froude numbers of these animals are equal when the camel is traveling three times as fast as the cat (the 3 squared in the dividend will cancel the 9 in the divisor).

Since I first suggested the possibility, many observations have confirmed that the movements of different-sized mammals are fairly nearly dynamically similar when the mammals are traveling with equal Froude numbers. For example, in dynamically similar movement, animals of different sizes would have equal relative stride lengths (stride length divided by leg length). Measurements from films confirm that a graph of relative stride length against Froude number is more or less the same for mammals as different as dogs, camels, and rhinoceroses—and even for bipeds such as people and kangaroos.

As the first of several insights we will owe to the Froude number, we will see how the speed at which gait changes depends on the size of the animal. Recall that the camel and the cat differed in their speed of movement at equal Froude number: the camel traveled three

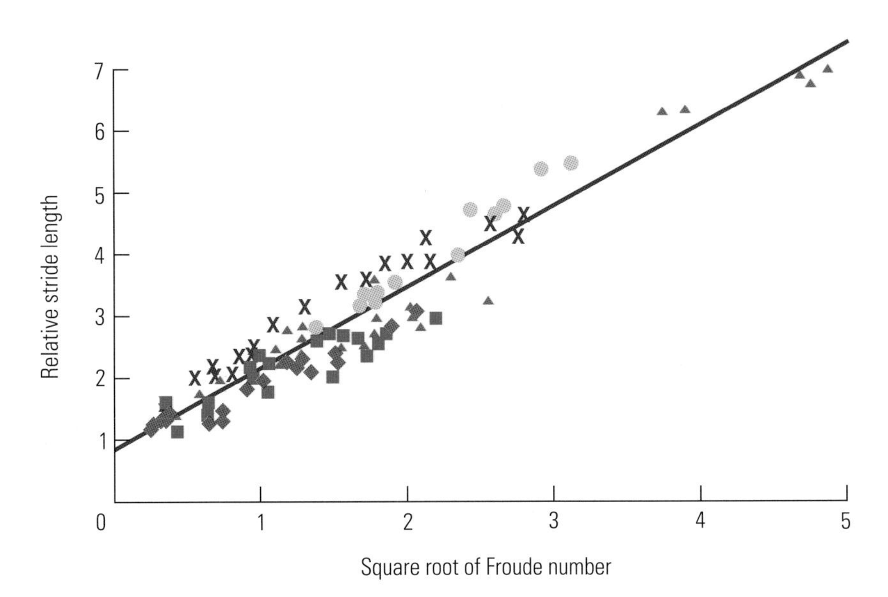


Animals of different sizes have about the same relative stride lengths when traveling with equal Froude numbers, as demonstrated by this graph of relative stride length plotted against the square root of the Froude number for × humans, ● kangaroos, ▲ dogs, ■ camels, and ■ rhinoceroses. The square root, rather than the Froude number itself, is plotted to avoid having the points clumped together too much on the left side of the graph.

times faster. We can expect a camel to change gaits at about three times the speed at which a cat makes the corresponding change. Speed measurements from films show that this is approximately true if you count the peculiar pace of camels as equivalent to the trot of cats. More generally, mammals change from walking to trotting or pacing at a Froude number of about 0.5 (1.0 meter per second for a cat, 2.9 for a camel) and from trotting to galloping at a Froude number of about 2.5 (2.2 meters per second for cats, 6.5 for camels).

This generalization fits well with a conclusion reached earlier in the chapter, that straight-legged walking is impossible when the speed $v$ is greater than the square root of the gravitational acceleration $g$ multiplied by leg length $l$—when $v$ is greater than $\sqrt{gl}$. This is equivalent to saying that stiff-legged walking is impossible at Froude numbers $v^2/gl$ greater than 1. The change from walking to running is actually made at a rather lower Froude number, but the example may help to show why Froude numbers are important.

## Energy Costs and Size

Although animals as different in size as cats and camels move in a similar manner, very tiny and very large mammals do not. If you look at the whole range of land mammals from shrews to elephants, you will see that their movements are not quite dynamically similar; smaller mammals run on bent legs and larger ones run with their legs much straighter. This has major consequences for the forces the muscles have to exert and for the energy they use.

The definition of dynamic similarity says that dynamically similar movements can be made identical by adjusting the scales of length, time, and *force:* in dynamically similar movements all forces are scaled up or down in the same proportion. An animal's weight is one of the forces that acts on it when it runs. Thus if different-sized animals moved in dynamically similar fashion, their muscles would exert forces proportional to their body masses. Each of the muscles of a 200-kilogram lion, for example, would have to exert 100 times as much force as the corresponding muscle of a 2-kilogram cat. The masses of geometrically similar animals are proportional to $(\text{length})^3$,

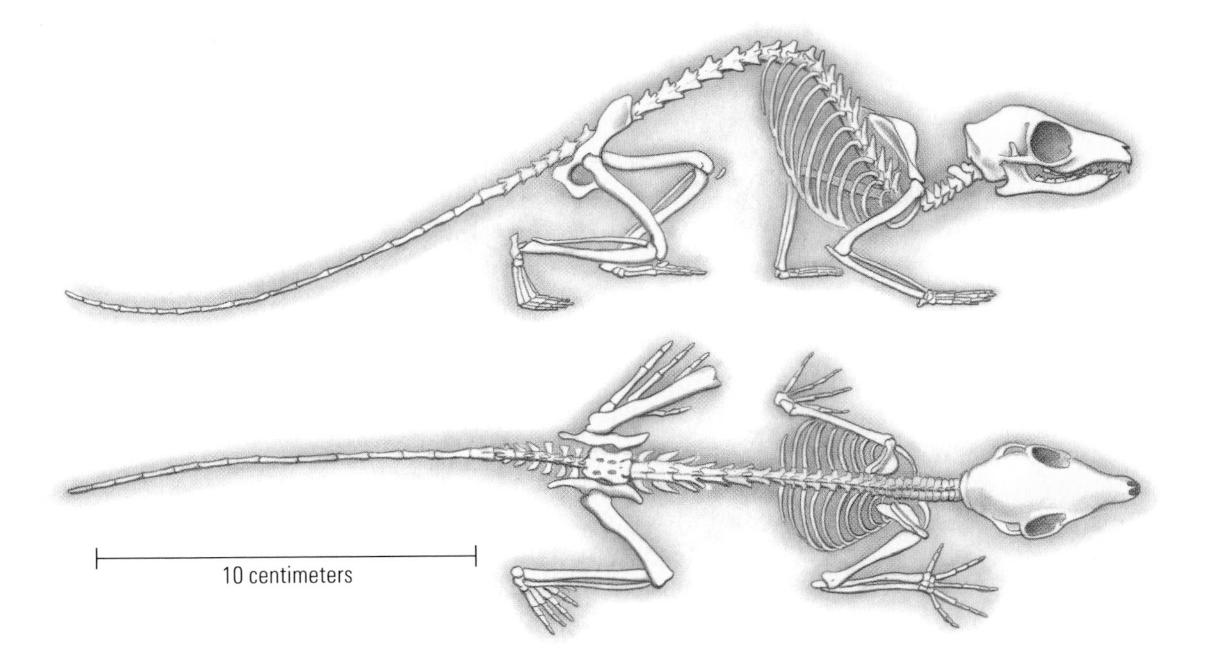


Small mammals such as this tree shrew run in a crouched position, with their legs strongly bent. This outline is based on X-ray pictures.

but the cross-sectional areas of their muscles are proportional only to $(\text{length})^2$, or $(\text{body mass})^{2/3}$. The cross-sectional areas of their muscles would be only $100^{2/3} = 22$ times as much in the lion as in a geometrically similar cat. We divide force by cross-sectional area to obtain the stresses in the muscles, which would be $100/100^{2/3} = 100^{1/3} = 4.6$ times as much in the lion as in the cat. That seems bad enough for lions, whose muscles would have to work much nearer their limits of strength than the muscles of cats have to do, but cats and lions are far from the extremes of mammal size. If a 3-gram shrew were scaled up to the size of a 3-tonne elephant and still ran on bent legs like a shrew, the stresses in its muscles would be increased 100 times.

Elephants and other large mammals avoid the need for impossible muscle stresses largely by keeping their legs much straighter than small mammals. (We saw when we discussed human walking how straight legs reduce muscle forces.) Andy Biewener of the University of Chicago has studied the posture of different-sized mammals and the dimensions of the muscles and their positions of attachment; he has concluded that when mammals ranging at least from 90-gram chipmunks to 300-kilogram horses use similar gaits, the peak stresses in their leg muscles are about the same.

It seems obvious that elephants cannot run on bent legs, but why don't shrews run on straighter ones? The forces in their muscles would be less, so they would use less metabolic energy. The most plausible reason suggested so far is that an animal with its legs bent is immediately ready to accelerate or jump, but one standing on straight legs cannot pounce or jump out of the way until it has first bent its legs.

Small mammals run on bent legs, and large ones keep their legs straighter; the difference is a matter of size. We are going to look now at the energy costs of running for different-sized animals, and we will learn that, weight for weight, large animals run more economically than small ones. Our attempts to explain that fact will eventually lead us back to the question of bent and straight legs. We begin by looking at measurements of oxygen consumption taken by Richard Taylor and his colleagues.

Richard Taylor's laboratory is an old missile site in the woods outside Cambridge, Massachusetts. There is plenty of room in the surrounding paddocks for the ponies that trotted or galloped on command in the gait experiments, and there is also room for more exotic animals. At various times Taylor has kept kangaroos, gazelles, cheetahs, and even young lions. Indoors he has moving belts to suit animals of all sizes, from 100-kilogram ponies down to tiny chipmunks less than a thousandth of that mass. There he and his colleagues have measured rates of oxygen consumption of a very wide range of mammals (and also of birds and lizards) walking and running at different speeds. They also shipped moving belts and oxygen analyzers to Kenya, where they measured mammals ranging from a 600-gram mongoose to a 240-kilogram eland (a large antelope). One of their most ambitious projects was to measure the oxygen consumption of a moving elephant, but even they were daunted by the prospect of building a moving belt strong enough to support the animal. Instead they made the measurements in a zoo while the elephant walked along paths and the oxygen analyzer traveled alongside on a golf cart.

Not surprisingly, this ambitious program of research showed that animals in general (like people and ponies) use oxygen faster when

Measuring the oxygen consumption of a walking elephant: an experiment by a team led by Richard Taylor.

running fast than when walking or running slowly. Also (again not surprisingly), large animals use oxygen faster than small ones traveling at the same speed. More interestingly, the research produced a general equation relating the energy used by a running animal to its speed and body mass.

We have already seen that graphs of metabolic rate against speed for people and ponies consist of a series of intersecting curves, each curve representing a different gait. If we are content to simplify the data by ignoring the curves, however, we can draw a single straight line showing energy consumption over the whole range of speeds. If we are not concerned with the relative merits of different gaits, that line is a reasonable representation of the data. It can be described by an equation,

$$P_v = P_0 + Cv$$

$P_v$ is the rate of consumption of metabolic energy when running at speed $v$ and $P_0$ is the rate of consumption when standing still. $Cv$ is the extra rate of energy use (over and above the standing rate) for running at speed $v$; therefore, C is the extra power (rate of energy use) divided by speed. In other words, C is the energy cost per unit distance:

$$C = \frac{\text{energy cost}}{\text{distance}}$$

C is a constant for each individual animal, but it is different for different species, especially for species of different sizes. You might expect locomotion to be twice as costly for a large animal as for a smaller one half its mass; in other words, you might expect $C/m$ to be constant (where $m$ is body mass). However, the real situation is more complicated. C is larger for larger animals, but not in proportion to their masses: $C/m$ is *smaller* for larger animals.

The graph on this page shows $C/m$ plotted against body mass. The scales have been made logarithmic so that the distance along the x-axis from 10 grams to 100 grams, for example, is the same as from 100 grams to 1 kilogram or even from 10 kilograms to 100 kilograms. Plotted in this way, the data points all lie close to a line that has a slope of $-0.32$. This tells us that $C/m$ is proportional to (body mass)$^{-0.32}$, so the cost of traveling a unit distance, $C$, is proportional to (body mass)$^{0.68}$. Every time you increase body mass by a factor of

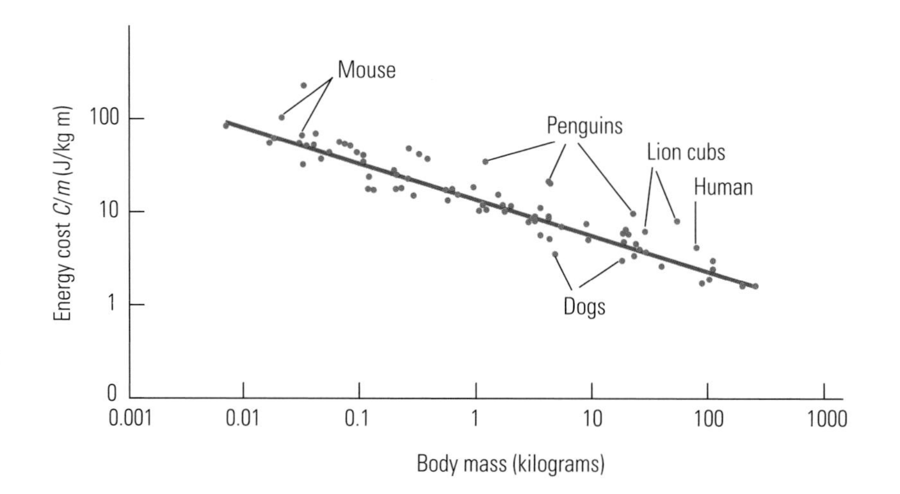


The energy cost of walking or running for various mammals, plotted against body mass. The labeled points show that mice, penguins, lion cubs, and humans have rather high energy costs for their sizes, and dogs rather low ones.

10, for example, you increase the energy cost per unit distance by a factor of $10^{0.68} = 4.8$. Some obviously ungainly animals such as penguins have a higher value of C than the line predicts, but the data are probably not accurate enough to say much about which animals move economically and which do not. Their main value is that they show the general relationship between C and body mass.

Can we explain that relationship? In Chapter 1 we saw that muscles use very little metabolic energy when they are resting, more when they are exerting a force without shortening, and more still if they shorten, doing work. Until recently scientists assumed that the main energy cost of walking and running was the cost of doing work. Let us see where that assumption led.

Imagine similar animals of different sizes, moving with dynamically similar gaits. The forces involved would be proportional to their body weights, and the distances covered would be proportional to their leg lengths. Thus the work (force times distance) done in corresponding movements would be proportional to body weight multiplied by leg length. The distance traveled in a stride would also be proportional to leg length, so work per unit distance would simply be proportional to body mass—that is, to $(\text{body mass})^{1.00}$. If muscles of different-sized animals worked with the same efficiency (and there is no very obvious reason why they should not), metabolic energy used per unit distance would be proportional to $(\text{body mass})^{1.00}$. As we have seen, it is actually approximately proportional to $(\text{body mass})^{0.68}$.

The discrepancy is serious because we are dealing with a very wide range of body masses. If energy cost were proportional to $(\text{body mass})^{1.00}$, a 1.5-tonne elephant would use 75,000 times as much energy per unit distance as a 20-gram mouse. The energy cost is actually proportional to $(\text{body mass})^{0.68}$, and the elephant uses only about 2000 times as much as the mouse. The different energy costs of running for different-sized animals cannot be explained simply in terms of the work that their muscles have to do.

In 1990 Richard Taylor and his student Rodger Kram offered a new explanation for the energy costs of running of different-sized animals: they assumed that the energy cost of exerting force is much more important than the cost of doing work. This assumption seems reasonably plausible, because runners need a lot of energy but may not do much work. Running at constant speed on level ground requires merely enough net work to overcome the air resistance and the friction in the joints. Nearly all the positive work done by muscles at some stages of the stride is cancelled out by negative work done at others. The metabolic energy consumption is increased while the muscles are shortening (doing positive work) and reduced while they are lengthening (doing negative work), but the total energy used in a complete stride may not be much more than if the muscles had exerted the same forces, without either lengthening or shortening. It may be possible to explain the energy cost of running almost entirely in terms of the cost of force production.

We saw in Chapter 1 that the energy cost of force production is expected to be

$$\text{cost of force} = \frac{\text{force} \times \text{fascicle length} \times \text{time}}{\text{economy}}$$

This equation tells us the energy used in a given time, but we would like to know the energy per unit distance, the quantity C that we have been discussing. Divide both sides of the equation by distance and remember that speed is distance/time:

$$\frac{\text{cost of force}}{\text{distance}} = \frac{\text{force} \times \text{fascicle length} \times \text{time}}{\text{economy} \times \text{distance}} = \frac{\text{force} \times \text{fascicle length}}{\text{economy} \times \text{speed}}$$

Kram and Taylor argued that the less time the feet stay on the ground at each footfall the faster (and so less economical) the muscles must be. For example, a mouse whose paws each stay on the ground for one twentieth of a second in each step needs faster, less economical muscles than a horse whose hooves each stay on the ground for half a second. They suggested that economy might be proportional to ground contact time:

$$\frac{\text{cost of force}}{\text{distance}} \text{ is proportional to } \frac{\text{force} \times \text{fascicle length}}{\text{ground contact time} \times \text{speed}}$$

Ground contact time multiplied by speed is the distance the animal's body travels while one particular foot is on the ground; this is called the step length:

$$\frac{\text{cost of force}}{\text{distance}} \text{ is proportional to } \frac{\text{force} \times \text{fascicle length}}{\text{step length}}$$

If animals of different sizes were geometrically similar to each other and used dynamically similar gaits, fascicle length and step length would be proportional to each other. As a consequence, the cost per unit distance of generating the required muscle forces would simply be proportional to the forces themselves. The forces would be proportional to body mass, and so the cost per unit distance would be proportional to body mass. That conclusion is identical to the one we reached when assuming that the cost of running was the cost of work. We do not seem to have made much progress toward understanding the energy costs of different-sized animals.

Animals of different sizes, however, do not use gaits that are precisely dynamically similar: small mammals like mice run on strongly bent legs, and large ones like elephants run on much straighter legs. We have already noted one consequence: peak stresses (force per unit cross-sectional area) are about the same in the leg muscles of mammals of widely different sizes. In geometrically similar animals, cross-sectional areas would be proportional to $(\text{length})^2$ or to $(\text{body mass})^{0.67}$. This argument suggests that the forces in leg muscles, and so the energy cost per unit distance ($C$), should be proportional to $(\text{body mass})^{0.67}$. This conclusion is almost exactly right: measurements of oxygen consumption already described show that C is approximately proportional to $(\text{body mass})^{0.68}$.

This explanation of the energy cost of running is impressive and very attractive, but we should remember that it is based on some doubtful assumptions. It assumes that the energy cost of running is predominantly the cost of force rather than of work, and it assumes that economy is precisely proportional to ground contact time. Our knowledge of muscle physiology is not yet good enough for us to be sure whether these assumptions are sound.

The Kram and Taylor theory seems to explain why many animals that run well have rather long legs for their body mass. For example,

Cheetahs *(Acinonyx jubatus)* are very fast runners, as are the antelopes they hunt, including the impala *(Aepyceros melampus)*, shown here. Impala often make spectacular leaps as they run.

a 60-kilogram gazelle has hind legs 0.80 meter long, whereas a 60-kilogram warthog has hind legs only 0.55 meter long. The longer legs of the better runners enable them to increase step length, and the theory tells us that the cost per unit distance should be proportional to 1/step length.

A similar argument provides a further explanation of the advantage of galloping. We have already seen how the work that the muscles have to do, swinging the legs backward and forward, is reduced in galloping by the spring action of an aponeurosis in the back. In addition, bending and extending the back enables the body's center of gravity to travel farther while a foot remains on the ground: it increases the step length and thereby (by our new argument) should reduce the energy cost of running. Greyhounds and cheetahs are fast runners with shorter legs than antelopes of the same mass, but with exceptionally flexible backs. This flexibility suggests that they should be economical runners, but measurements of oxygen consumption while running do not show much difference between cheetahs and other mammals of similar size.

Antelopes, horses, greyhounds, and cheetahs seem to be among the fastest running animals, but again it is difficult to be sure. Many of the animal running speeds that can be found in books are merely subjective impressions based on the observer's experience of road traffic. Others are speedometer readings made by driving alongside a running animal in a vehicle, but they too are often unreliable. For example, if the animal tries to swerve away from the vehicle it will be traveling on the inside of a bend and the vehicle on the outside: the vehicle therefore must travel faster than the animal to keep alongside. The only large animals for which maximum speeds have been reliably measured are greyhounds and racehorses. Times given in the sporting pages of newspapers show that most greyhound races are won at 15 to 16 meters per second (34 to 36 miles per hour) and that most horse races are won at 16 to 17 meters per second (36 to 38 miles per hour). These animals have been bred for speed and seem likely to be faster than most wild animals, which have been selected for in the course of evolution not only for speed, but also for other qualities. With colleagues, I have filmed many of the larger Kenyan mammals running in their natural habitat, pursued by a vehicle. The highest speed shown in our films is only about 14 meters per second, both for Thomson's gazelle and for zebra. People are even slower. A good athlete who runs 100 meters in 10 seconds is averaging only 10 meters per second, and the peak speeds of the world's best sprinters are only about 12 meters per second.

Many books will tell you that cheetahs can run at 70 miles per hour (31 meters per second). Such claims seem to be based on a popular article that described a tame cheetah running 80 yards in $2\frac{1}{4}$ seconds. Unfortunately, according to a later article, the enclosure where the test was made is only 65 yards long, so the speed that is claimed is probably much too high. A later record of 56 miles per hour (25 meters per second) is still astonishingly high, although more believable.

## Sprawling Gaits

Birds and mammals run with their feet close under the body, so the lines of footprints made by their left and right feet are close together. Fossil footprints of dinosaurs show that they walked in the same way, but modern reptiles run with their feet well out on either side of the body, so that the lines of left and right footprints are well apart. This

Lizards use a sprawling gait and bend their backs from side to side as they run. This gecko is climbing a cactus plant.

manner of running is usually described as a "sprawling" gait. Reptiles move diagonally opposite feet together, the left fore with the right hind and the right fore with the left hind. This sequence is the same as in mammalian trotting, but lizards also use it for slow gaits. They bend their bodies from side to side as they run, timing the bending so that it increases the length of their steps.

Until recently, zoologists assumed that the sprawling style of running was uneconomical because it required large forces in muscles. (The legs are held much as people hold their arms in a press-ups exercise.) We should have realized that, though muscle forces may be high, they are also high in small mammals such as rats that run on bent legs. Only when the rates of oxygen consumption of running lizards were measured was it realized that they run as economically as mammals. The energy cost per unit distance (C in the equation on page 43) is about the same as for mammals of equal mass, and the metabolic rate while standing still ($P_0$) is considerably less than for mammals. Why then did posture change in the course of evolution, from the sprawling stance of early reptiles to the mammalian stance with the feet tucked in under the body?

An important clue emerged when films of lizards were analyzed and it was observed that they run by fits and starts, stopping briefly between bursts of about 2 to 12 strides. Dave Carrier of the University of Michigan showed that they stop to breathe; running and breathing use the same muscles in different ways, and so the two activities cannot be performed simultaneously. In running, the side-to-side movements of the body require the muscles of the left and right sides of the trunk to contract alternately. In breathing, however, the muscles of the two sides must act together to enlarge the rib cage and draw air into the lungs, or to compress it and drive air out. Mammals have no such difficulty: indeed, the movements of galloping seem to help breathing. The mammalian back, bending up and down instead of from side to side, works like a bellows. When it bends it squeezes the body cavity, driving air out of the lungs, and when it straightens it draws air in. Records of the flow of air through galloping horses' nostrils show that they take one breath per stride, breathing out as the back bends and in as it extends again.

The sprawling gait can be fast as well as economical. As a general rule, lizards can sprint about as fast as mammals of equal mass. Quite small (50-gram) lizards have been timed at the remarkably high speed of 8 meters per second (18 miles per hour).

In contrast, tortoises are notoriously slow. We have already seen that their slow muscles are very economical, but now we will consider how slowness affects stability. If you try to ride a bicycle very slowly, you will probably wobble and fall off. Similarly, walking very slowly requires more precise control of the forces on the ground than walking faster or running. The reasons are not the same as for slow bicycling (there is nothing in walking comparable to the stabilizing gyroscope action of bicycle wheels), but the effect is the same: slow movement requires more precise control. I will try to explain why.

Whatever the speed of walking, the feet exert fluctuating forces on the ground, and the body is generally not in equilibrium. At one stage the forces on the feet may total more than body weight and accelerate the body upward, but at another they may be smaller and let the body fall. The body will rise and fall in the course of a stride but this may not matter, unless the vertical movements are so big that the belly bangs on the ground. Similarly, temporarily unbalanced horizontal forces may make the animal speed up and slow down during each stride, or veer from side to side. Imbalance between the forces exerted by different legs may tilt the animal, causing it to pitch and roll. We may think of the rising and falling, speeding and slowing, pitching and rolling as unwanted movements, but walking remains effective if they are not too large.

When an animal goes fast, footfalls follow each other in rapid succession, and an unwanted movement started by an unbalanced force at one stage of a stride can soon be corrected. If the animal is moving slowly, however, the unwanted movement will continue for longer before there is an opportunity to correct it, and it may go too far. If the animal is moving very slowly, force fluctuations must be kept small.

The problem is especially severe for a low-slung animal, such as a tortoise that supports its shell close above the ground. A small loss of height or a tilt through a small angle may make it hit the ground, which presumably would be unsatisfactory. A quick calculation will make the problem more obvious. Imagine that an animal's legs suddenly stopped exerting any force, so that the trunk was completely unsupported. If the animal were a tortoise with its belly initially 5 centimeters from the ground, it would hit the ground after only 0.10 second. For a tortoise of this size, each stride would last about 2 seconds, so the shell would hit the ground if the animal was unsupported for a mere twentieth of a stride. However, if the animal were a dog with its belly 40 centimeters from the ground, it would take 0.28 second to fall. The dog's strides would each last about 0.5 second, so in its case the falling time would be half a stride period. The tortoise could tolerate only a tiny unsupported fraction of a stride (or smaller force fluctuations for larger fractions), but the dog's feet could be off the ground for quite a large fraction of a stride without ill effects.

In theory it is possible for a tortoise or other quadruped to keep perfect equilibrium through each stride without rising, falling, pitching, or rolling at all. Three feet are enough to support a body in stable equilibrium: a three-legged stool is stable, but a two-legged one is unstable and will fall over. The tortoise could remain stable by moving one foot at a time, always leaving three on the ground for support.

If you sit on a three-legged stool and lean over too far to one side, it will fall over. All is well so long as the center of gravity (of you and the stool combined) is vertically over the "triangle of support" formed by the stool's three feet, but if the center of gravity moves outside the triangle, you will topple. Similarly, an animal standing on three feet cannot remain in equilibrium if its center of gravity is not over the triangle of support. The problem disappears if the feet are big enough and can be placed directly under the center of gravity (birds and people can stand on one foot), but animals with small feet need three, suitably placed.

It turns out that it is not enough for a quadruped to move one foot at a time: to keep the center of gravity always over the triangle of support, it must move its feet in a particular order. This order, shown in the diagram on this page, is actually used by dogs, horses, and most other quadrupeds whenever they walk. Tortoises, which have far more need to keep close to equilibrium, also move their feet in this order, but unlike large mammals they do not move their feet at equal intervals. Each fore foot is set down only slightly before the diagonally opposite hind foot, so the footfall pattern is very like that of lizards (which move diagonally opposite feet together), and there are times when only two feet are on the ground.

At first sight this gait seems faulty, but Alan Jayes and I were able to show that, for an animal with slow-acting leg muscles, it would allow steadier walking than if the feet moved one at a time. The reason is that the "ideal" gait in the diagram can keep the animal in

If a tortoise moved one foot at a time in the order shown here, its center of gravity would always be over the triangle of support provided by the feet on the ground.

constant equilibrium only if the forces exerted by the other feet can be changed instantaneously, whenever a foot is lifted or set down. If the muscles are incapable of rapid adjustments, the animal can keep closer to equilibrium by moving its feet in diagonally opposite pairs.

Keeping close to equilibrium is easier with six or more legs. Insects generally move their six legs in two groups of three, the two groups taking turns to provide the triangle of support. Each group consists of the front and rear legs of one side of the body and the middle leg of the other.

Centipedes may have to use gaits that keep their many legs out of one another's way. *Scolopendra* is a short-legged centipede that moves the legs of each side of the body in sequence from front to back: leg 1 before leg 2 before leg 3, and so on. In a group of adjacent feet that are on the ground, the ones in front are at a later stage of the step than the ones behind, so the feet on the ground form clumps, as the diagram on this page shows. This gait presents no problem for *Scolopendra*, but if the legs were much longer they would cross over each other so that the foot of leg 2 was set down in front of foot 1, and

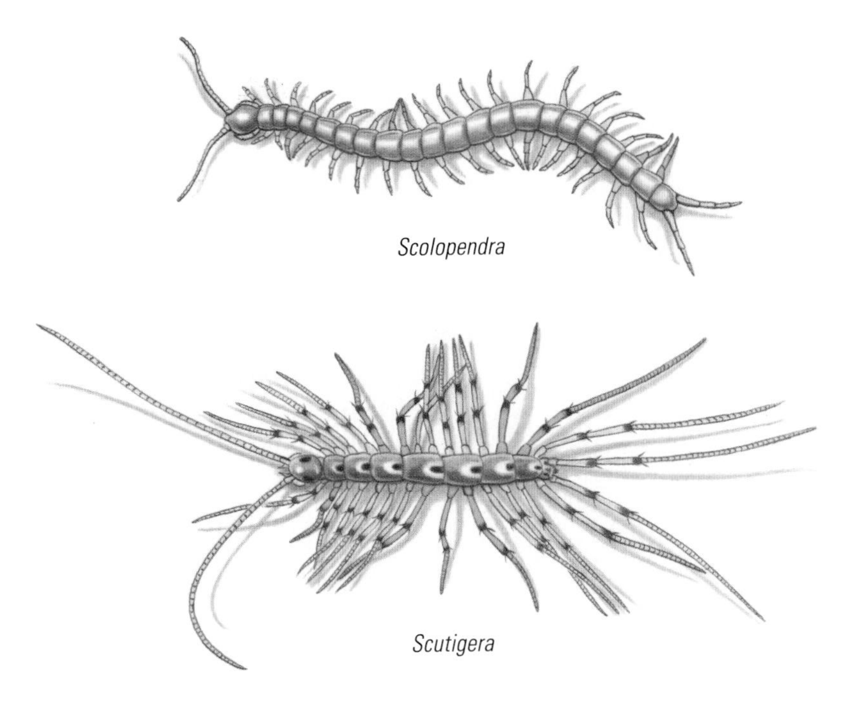


*Scolopendra,* a short-legged centipede, moves its legs in sequence from front to back; *Scutigera,* with long legs, moves them in the reverse sequence. Legs whose feet are on the ground are tinted.

A walking millipede's legs form a regularly repeating pattern because each moves slightly after the one behind. Whereas most walking animals move the left and right legs of a pair alternately, millipedes move them together.

foot 3 in front of foot 2. It is hard to see how the centipede could do that without getting into a tangle. Another diagram shows how *Scutigera,* a long-legged centipede, avoids the difficulty. It moves its feet in reverse sequence, starting at the rear. The result is that the feet are spread out instead of being clumped or crossed.

Insects, centipedes, and other arthropods use sprawling gaits, keeping their feet well out on either side of the body. For reasonably large land animals, sprawling gaits are optional: lizards and crabs use sprawling gaits, but birds or mammals of similar mass do not. For small animals such as insects, sprawling is probably essential: if they did not stand with their feet well out to either side of the body, they would be apt to be blown over by gentle breezes. The reason is that the force of the wind on an animal is proportional to its surface area, but the weight that helps to stabilize the animal is proportional to its volume. For geometrically similar animals of different sizes, smaller ones have larger ratios of area to volume and so are more likely to be blown over.

This last section has pointed to the importance of stability in the evolution of the gaits of tortoises and insects, but most of the other messages of the chapter have been about energy. Walking or running, trotting or galloping, animals use gaits that have been adapted to save energy, whether by the principle of the pendulum or by the bouncing of tendon springs. And the speeds at which they change gaits are the speeds at which the faster gait becomes more economical of energy than the slower one. The broad picture seems clear, but some of the details are hazy because our understanding of muscle physiology is still imperfect. We are not at all certain about the relative importance of the two kinds of energy costs, those of the forces that the muscles exert and those of the work that the muscles do.

# 3

# Climbing, Jumping, and Crawling

Grasshoppers could not jump so well if they did not have a catapult mechanism built into their big hind legs. This enables them to extend the legs much faster than the muscles could do unaided.

The previous chapter was about walking and running, the standard methods of traveling on legs. This one is about other ways of moving over solid surfaces, some using legs and others not. We will discover how built-in catapults help fleas to jump; why cats that can climb a tree easily may have difficulty getting down again; how flies manage to walk on the ceiling; and why crawling snails leave a trail of slime behind them.

## Jumping

Animals jump for many different reasons. Lions jump onto prey to bring them down. Frogs and grasshoppers jump in directions that are hard to predict, to escape from danger. Squirrels and bushbabies jump between the branches of trees. Fleas resting on leaves jump when they sense the warmth of the body of a passing mammal or bird: with luck, they may be able to attach themselves to the potential host.

The secret of a good jump is speed. A fast takeoff will carry an animal closer to its target or farther from its enemy. As we will see, it turns out that the smaller the animal, the more difficult it is to reach a good jumping speed by the action of muscle alone. The muscles of smaller creatures need an assist, and for this reason we will find the most varied, and the most bizarre, jumping mechanisms in insects. Indeed, we will see one type of beetle that manages to jump without even using its legs.

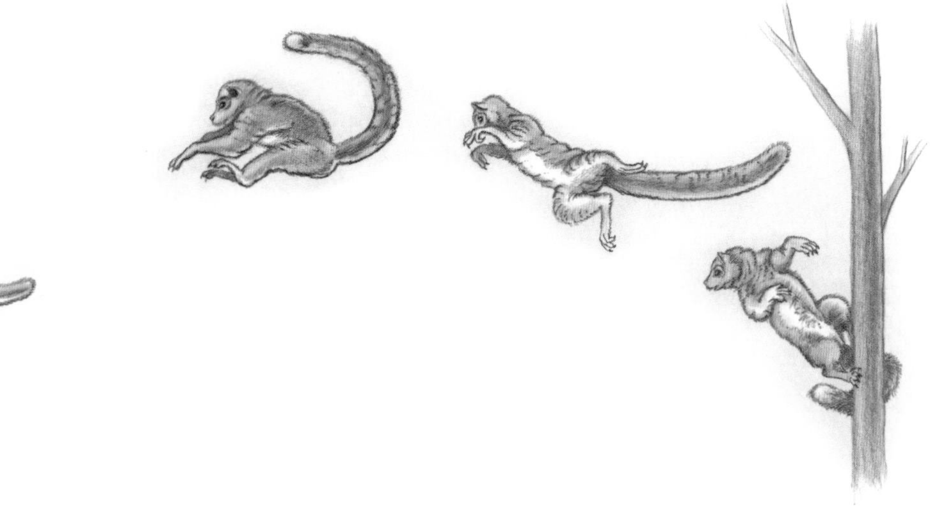

Bushbabies (this one is *Galago senegalensis*) leap between branches in their forest habitat.

Whatever the function of the jump, there is a simple relationship between the speed and angle at which the animal takes off and the distance it covers or the height it clears. It is particularly easy to calculate the height of a vertical jump. The kinetic energy of a body of mass $m$ traveling at speed $v$ is $\frac{1}{2}mv^2$. If the body rises through a height $h$, it gains potential energy $mgh$ ($g$ is the gravitational acceleration). In a vertical jump all the kinetic energy at takeoff is converted into potential energy at the top of the jump:

$$mgh = \tfrac{1}{2}mv^2$$

The height is therefore

$$h = \frac{v^2}{2g}$$

In these equations, $h$ is the increase in height of the body's center of gravity from the instant of takeoff to the top of the jump. For example, in a standing jump a human athlete might leave the ground with a vertical velocity of 3 meters per second, and the center of gravity would rise about $3^2/(2 \times 10) = 0.45$ meter. (The gravitational acceleration is about 10 meters per second squared.) The body's center of gravity is a little above the hips, so it would be about 1 meter from the ground at takeoff and would rise to 1.45 meters at the top of the jump. Senegal bushbabies are very much smaller than we are, weighing only 250 grams, but these tree-dwelling primates are very much better jumpers. A pet bushbaby was once observed jumping from the floor to the top of a 2.26-meter door. Its center of gravity could have been no more than 0.25 meter from the floor at takeoff, so the height gain $h$ was 2 meters, which requires a takeoff speed of more than 6 meters per second.

People often rate fleas as the most remarkable of animal jumpers. Though they are so small, fleas can jump heights of at least 130

A jumping bushbaby (*Galago senegalensis*) straightens its legs rapidly in order to throw itself into the air.

millimeters (5 inches). In absolute values their jumps are low in comparison to human jumps, but relative to body length they are much higher. Excluding the legs, a flea's body is only about 2 millimeters long and a man's body (again excluding the legs) about 1 meter. Therefore it might be argued that a flea's 130-millimeter jump is equivalent to an impossible 65-meter human jump.

The fallacy of that argument was pointed out in 1950 in an article by the famous muscle physiologist A. V. Hill. Hill wanted to figure out how the height of a jump depended on the size of a jumper. He would then be able to compare the expected jump heights for different-sized animals, including a flea and a man. He first argued that the work available for a jump should be proportional to body mass. The work done in a single muscle contraction is the force multiplied by the shortening distance. That force is proportional to the cross-sectional areas of the muscles, and we can expect the muscles to be able to shorten by amounts proportional to their initial lengths. Therefore, the work done in a single contraction is proportional to the cross-sectional area multiplied by the length—that is, to the volume of the muscle. If the animals being compared have equal proportions of muscle in their bodies, we can expect the work available for a jump to be proportional to body mass.

The work required for a jump to height $h$ equals the potential energy gain, $mgh$. If this work is proportional to body mass $m$ in animals of different sizes, all should be able to jump to the same height $h$. The flea that raises its center of mass 0.13 meter in a jump is doing much less well than a human who rises 0.45 meter. We should not ask why fleas jump so high, but why their jumps are so feeble.

Hill's theory said that different-sized animals should jump to equal heights, but ignored a problem faced by very small jumpers. People, bushbabies, and fleas all prepare for a jump by bending their legs. To take off, they extend the legs rapidly, accelerating over a distance of about 0.4 meter for people, 0.16 meter for bushbabies, and only about 0.5 millimeter for fleas. Humans accelerate from rest to about 3 meters per second at takeoff, so the average speed, as the legs extend, is about 1.5 meters per second; thus the 0.4-meter acceleration distance is covered in about 0.27 second. Films confirm that this estimate is about right. The simple equation on page 59 tells us that a flea jumping to a height of 0.13 meter would have to take off at

1.6 meters per second. (It would actually have to take off rather faster because the equation ignores the effect of air resistance, which slows small jumpers much more than large ones.) As the flea accelerates during takeoff from rest to over 1.6 meters per second, its *average* speed would be a little over 0.8 meter per second: it would cover the 0.5-millimeter takeoff distance in only 0.6 millisecond.

No muscle can complete a single contraction in so short a time. I was careful to write "single" because some midges beat their wings at 1000 cycles per second; a midge has only 0.5 millisecond for each up or down stroke. Such frequencies depend on the wings and their muscles going into a state of resonant oscillation, as will be explained in Chapter 5. A jump requires a single contraction to extend the legs, and no known muscle could do that in as little as 0.6 millisecond.

The flea's jump is possible only because the animal has a built-in catapult, a tiny block of rubberlike protein at the base of each hind leg. These blocks are too small for mechanical testing but seem to be made of a protein called resilin that is also found in the tendons of some dragonfly wing muscles and in the hinges that join locust wings to the body. Tests on these larger (but still very small!) pieces of resilin show that its properties are very like those of soft rubber. It can be stretched to three times its initial length and returns nearly all the energy in its elastic recoil.

A flea jumps from the fur of its weasel host.

When children use catapults they first stretch the rubber, storing up elastic strain energy in it. If the catapult is a strong one the child may be unable to stretch it fast, but a slow stretch will store the elastic strain energy equally well. When the catapult is released, the rubber recoils very rapidly indeed, projecting the stone much faster than it could have been thrown. Work done slowly on a catapult is returned much faster. Fleas store up strain energy in their resilin by a relatively slow contraction of their leg muscles lasting, it seems, about 100 milliseconds. A trigger mechanism releases the resilin, which recoils, extending the legs in only 0.6 millisecond and throwing the insect into the air.

The details of the mechanism are hard to make out, and zoologists disagree about how it works. We have a much better understanding of the jumping of locusts, which comes from the research of Dr. Henry Bennet-Clark of Oxford University.

The strong hind legs of a locust launch it into the air at a speed of up to 3.2 meters per second, fast enough to propel the insect about

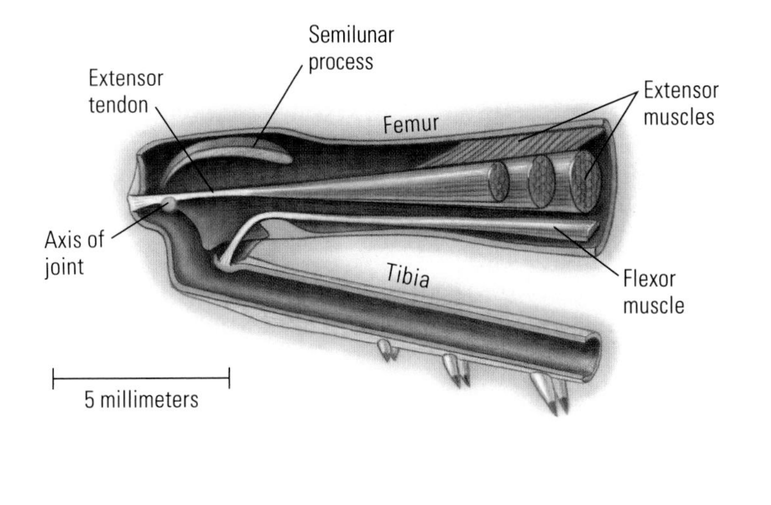

*(Left)* The springs in a locust's leg. *(Right)* In this desert locust, *Schistocerca gregaria,* a semilunar process (one of the springs that aid jumping) can be seen as a dark crescent at the "knee" joint of the hind leg.

0.5 meter (20 inches) if it jumped vertically. In fact locusts jump at an angle and rise less high but can clear horizontal distances of about 1 meter. Like fleas, they must rely on elastic recoil, but instead of using a tiny block of resilin as a catapult they store elastic strain energy by means of a tug-of-war between two muscles.

About halfway along the locust leg is a joint called the knee, by analogy with the human leg. Kept bent when the animal is standing, the joint extends very rapidly to throw the animal into the air. A large extensor muscle filling most of the locust's fat thigh extends the knee and does most of the work needed to power the jump. A much smaller flexor muscle bends the knee and prevents the joint from extending too soon. Like many other insect muscles, these are pennate; the fibers converge on a central tendon made up of material very like the exoskeleton that encloses the whole body. When the insect is preparing for a jump, both muscles contract. The extensor is much the stronger but, while the knee is bent, its tendon runs much closer to the axis of the joint than does the tendon of the flexor. This difference of lever arms enables the relatively weak flexor to hold the knee bent, against the stronger pull of the extensor. The large forces that are developed at this preparatory stage compress the semilunar processes (flexible parts of the leg skeleton, close above the knee) and stretch the extensor tendon, storing elastic strain energy in both. To jump, the animal suddenly relaxes the flexor muscle, allowing the

Click beetles, which fall upside down onto the ground, jump by a jackknife action of the back. The bent lines above the insect are outlines of the insect's belly traced from successive frames of a high-speed film.

knee to extend, which it does very rapidly, driven by the elastic recoil of the semilunar processes and the extensor tendon.

One of the most curious of jumping insects is the click beetle, which can jump twice as high as a flea without so much as bending a leg. The click beetle flexes its back instead. These beetles are often found crawling on blades of grass, but drop to the ground if disturbed. If they land upside down and cannot right themselves quickly, they jump by a sudden jackknife movement of the back that may throw them as high as 0.3 meter (12 inches) off the ground. The jump has been filmed using a special camera running at 3100 frames per second. (Ordinary films are taken at only 18 to 24 frames per second.) Even at this exceedingly high rate, the films showed the beetles accelerating to takeoff in just two frames, or 0.6 millisecond. As in flea and locust jumping, the rapid acceleration of a click beetle jump is made possible by a catapult mechanism.

All these insects take off so quickly that there is scarcely time for muscles to start shortening during takeoff; the jumps are almost entirely powered by elastic recoil. Larger jumpers such as bushbabies and human beings have time to shorten their muscles during takeoff, but these more muscle-dependent jumpers also benefit from tendon elasticity. If you ask people to do a standing jump, they will generally bend their knees and then immediately extend them. If they pause with knees bent and then jump, they cannot jump so high. The reason is that muscles can exert more force while being stretched than while shortening or holding constant length. The muscles that extend the legs at takeoff also stop the downward preparatory movement; as they stretch to counteract the bending of the knee, they are able to develop large forces. These forces stretch the tendons in turn, storing strain energy that is returned in elastic recoil, increasing the height of the jump.

## Swinging Through Trees

Squirrels and many monkeys traveling through the forest run along the tops of horizontal branches much as they run on the ground, but leap as necessary from one branch to another, often moving from tree to tree. Bushbabies and lemurs move through trees in a different way, clinging to vertical branches and leaping from one to the next. But some tree dwellers have found an efficient way to move through the forest while barely using their legs at all—these animals propel themselves by the strength of their arms.

Apes and a few monkeys swing by their arms below branches, using the technique called brachiation. Particularly good at brachiation are gibbons, Southeast Asian apes with very long arms and relatively short bodies and legs. Gibbons often have to travel along slender branches to reach fruit. Their method of travel is especially effective for traversing a slender branch, because walking on top of a thin branch is a feat of balance like tightrope walking, but swinging below a branch presents no such problems.

Children make swings go higher by "pumping"—bending and extending their legs at appropriate stages of the swing. Gibbons make use of the same technique, but instead of using it to go higher, they speed up their swinging through the trees. As they swing under a

A brachiating siamang (*Hylobates syndactylus*, a species of gibbon) builds up speed like a child "pumping" a swing.

Orangutans *(Pongo pygmaeus),* like this juvenile, swing through trees much as gibbons do.

branch they bend their legs, raising their potential energy. At the top of the swing, as they reach out to catch the next branch, they extend their legs again. If at this stage they are flying through the air between branches, the extending of the legs lowers the feet and raises the trunk a little but leaves the path of the center of gravity unaltered. And if the center of gravity's path is unchanged, the body's energy is unchanged as well. Bending the legs increases the body's energy and extending them leaves it unchanged, so the effect of the whole cycle of pumping movements is to increase the energy, making the animal travel faster.

A child pumping a swing stands up as it passes through the vertical and bends the knees at the top of the swing. This stand-and-bend sequence seems the reverse of the gibbon's action if you look only at the legs: the child is straightening its legs when the rope is vertical, but the gibbon is bending its legs when its arm is vertical. However, the effect on the center of mass is the same in both cases: it is being raised at the stage when rope or arm is vertical.

*(Top)* The forces on a squirrel climbing a tree. *(Bottom)* Squirrels (such as these five-striped palm squirrels, *Funambulus pennantii,* from India) have mobile ankles that enable them to reverse their hind feet for climbing down trees.

## Gripping a Smooth Surface

Gibbons spend their lives in the trees, seldom or never descending. In contrast, squirrels and many monkeys often come down to the ground, where they find much of their food. To get back up a tree, these animals need to climb a vertical trunk. In accomplishing their climb, not only must they be correctly balanced to counteract the force of gravity, but they need a way to hang on to the surface. Indeed, for some animals, the difficulty of maintaining a good grip can make getting down trickier than getting up. As adroit as climbing mammals are at maneuvering up and down trees, however, the true masters are some insects and lizards that are able to hold on to smooth surfaces, even when upside down.

To climb a vertical trunk, an animal must pull with its fore limbs and push with the hind. The forces exerted by the limbs must balance the animal's weight, so you might think that only vertical forces would be needed. But if the only forces on the animal were its weight acting downward and upward forces on its paws, it would not be in equilibrium: that combination of forces would make it rotate head

This kinkajou's *(Potos flavus)* ankles can be reversed to enable it to hang by its hind feet.

over heels. For equilibrium, the forces on the feet must slope, as shown in the diagram on page 67. The force on the fore feet may be more vertical and the force on the hind more horizontal, or vice versa. The (downward) weight is then balanced by the (upward) components of the forces on the paws; the (forward) component of force on the fore paws is balanced by the (backward) component on the hind paws; and forces that would rotate the animal clockwise are balanced by others that would rotate it counterclockwise.

Squirrels have claws that dig into bark and help them climb. The hind claws are pushed into the bark, so the animal does not have to rely on friction to prevent it from sliding down the tree. The fore claws are hooked into the bark, enabling it to pull on them. Claws on the hind feet are not essential for this kind of climbing unless the surface is slippery, but claws on the fore feet are essential. The squirrel could not pull on its fore feet if its claws were not dug in.

The opposite is true for coming down a tree trunk headfirst: the hind claws must be hooked in. Not only does the animal need hind claws, but it also must be able to turn its feet back to front so that the claws can pull in the required direction. Squirrels can do this, and so can kinkajous and various other mammals that live in trees. Domestic cats cannot reverse their hind feet, though the South American tree-living cats called margays can. A domestic cat that climbs up a tree easily may have great difficulty coming down.

Reversing the hind foot as squirrels and kinkajous can do requires a very mobile ankle. To understand what is required, sit on the floor

This nuthatch (*Sitta europaea*) has clawed toes facing forward and back that enable it to climb up or down tree trunks.

with the soles of your feet flat on the floor in front of you. You will be able to turn your feet inward so that the soles of the feet are vertical, each facing the other. To do as squirrels and kinkajous do, you would have to be able to turn them through a further 90 degrees until the soles were horizontal, facing upward.

Coming down tree trunks is relatively easy for birds because their gripping feet have one or more backward-pointing toes, so there are claws curving in the right direction to hook into the bark whether the bird is running up or down the tree. Creepers (known as tree creepers in Britain) work their way up tree trunks, searching for insects in the crevices in the bark, then fly to get down again, but nuthatches run down tree trunks as well as up. Creepers and woodpeckers have stiff tail feathers, which they press against the tree trunk, using them to supply the force that climbing squirrels get from their hind feet.

Only one group of small monkeys, the marmosets, have claws, and even they lack a claw on the big toe. Other monkeys have fingernails and toenails, like humans, and rely on friction for their grip. To be able to pull with their arms (which they must do to climb vertical trunks), they must be able to reach well around toward the back of the tree. They need long arms to be able to climb even moderately large tree trunks, and they cannot climb very thick ones.

The adhesive feet of insects such as this stink bug (suborder Heteroptera) enable them to walk on the surfaces of leaves.

The accomplishments of monkeys pale beside those of creatures that seem able to climb the smoothest surfaces and even defy gravity by walking upside down. Flies and other insects can walk up vertical walls and even windowpanes that give no chance for claws to hook in. They can also walk upside down on the ceiling. Small lizards called geckoes are often seen hanging upside down on the ceilings of houses in the tropics, and tree frogs can cling to vertical surfaces by means of the soft pads on their feet. All these animals depend on adhesive feet.

Aphids (greenfly and similar bugs) suck out the juices from plants through mouthparts that resemble a hypodermic needle. They have to walk on smooth plant surfaces, including vertical stems and the undersides of leaves, and they have to adhere strongly enough not to be blown off or shaken off as the leaf waves in the wind. The force of adhesion has been measured by putting an aphid on a clean piece of glass on the pan of a sensitive scientific balance. The balance is adjusted to read zero with the glass and aphid in place, then a thread that has been dipped in glue is dangled over the aphid so as to stick to it. The investigator pulls gently upward on the thread, making the balance register a negative load. The balance reading at the instant when the aphid detaches is about −6 milligrams (thousandths of a gram) for aphids weighing 0.3 milligram. These aphids can hold on to glass with a force 20 times their own weight.

That experiment measured the force at right angles to the glass needed to detach the aphid, but aphids can resist forces at least as large when they act parallel to the glass surface, as a gust of wind might. This has been demonstrated by letting aphids crawl up the sides of glass centrifuge tubes and then spinning them at various speeds and observing whether they were thrown to the bottom of the tube.

Aphids have tiny claws on their feet, but you would not expect these to be any use for attaching to smooth glass surfaces. Thus it is not surprising that aphids can still adhere after the claws have been cut off with a fine scalpel. The attachment organ is a spongy pad on the foot. Various experiments have been performed to find out how it works.

One possibility is that the pads attach by suction. If there were muscles that could lift the center of the pad, the pressure under the pad would be reduced and it would hold on like a sucker. However, no likely muscle has been found, and the force of adhesion is affected very little by anesthetizing the aphid with carbon dioxide, which makes muscles relax. Furthermore, and this is the conclusive evidence, aphids remain firmly attached in a vacuum.

An important clue comes from the observation that aphids lose the ability to adhere to smooth surfaces after walking for a while on silica gel, which absorbs water strongly. Aphids that had walked on silica gel for 15 minutes took about 30 minutes to recover their power of adhesion—unless they were placed on water-soaked filter paper. Allowed to walk on this moist surface, they recovered within 2 minutes. Adhesion seems to depend on water.

The diagram at right shows how the mechanism probably works. There is a thin film of water between the foot and the surface to which it is attached. Pulling the foot away from the surface, widening the gap, would draw the water surface in around the edges of the foot. But because this inward pull is resisted by surface tension, a large negative pressure can develop in the water. The negative pressure holds the aphid in place.

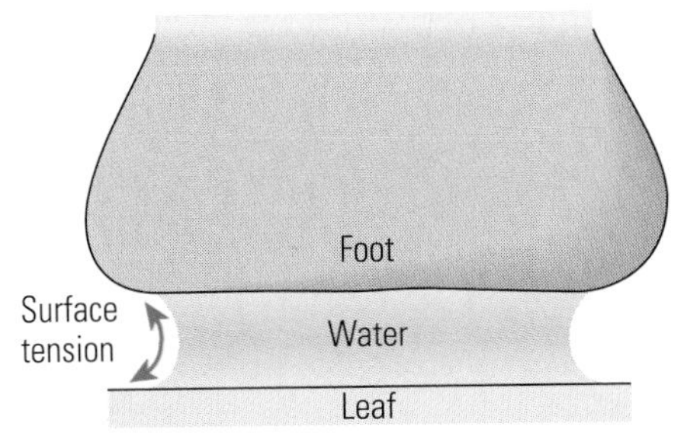


Surface tension allows negative pressures to develop in the film of water between an aphid's foot and the surface on which it is standing.

*(Left)* A scanning electron micrograph of a foot of a beetle *(Bruchus atomarius)* showing the setae (bristles) that give it a grip on smooth surfaces. *(Top right)* This mottled gecko (*Phyllodactylus marmoratus,* from Australia) is climbing a pane of glass. *(Bottom right)* A scanning electron micrograph of part of a toe pad of another species of gecko shows the branched setae that give it its grip.

The feet of beetles and some other wall-climbing insects are very different from those of aphids. Instead of having spongy pads, their feet are covered by a dense pile of very fine, flexible setae (bristles) with wider ends. Such setae can make exceedingly close contact with any solid surface, whether it be smooth or rough, and it has been suggested that they attach by van der Waals forces, which are forces of attraction between molecules. These forces are very weak unless the adhering surfaces are very close together: the force is inversely proportional to the cube of the distance, so if you move twice as close you get eight times the force. For insect adhesion to work this way, adhering surfaces would have to be no more than about 10 nanometers, or about the diameter of a large protein molecule, apart. (A nanometer is one millionth of a millimeter.) There seems to be no direct evidence that van der Waals forces are involved, but all the other obvious possibilities have been fairly convincingly eliminated: no liquid or adhesive has been observed, beetles remain attached in a vacuum, and they are not detached by an antistatic gun, which would release them if they depended on electrostatic forces.

Geckoes have setae on their feet like those of beetles, only finer. Even the ends, where they widen, are only about 200 nanometers across, and at this width the setae are too narrow to be visible by light microscopy. Electron microscopy is convenient for looking at the setae on beetle feet and is essential for observing the setae of geckoes.

## Traveling Waves

We walk on legs and think of that as the normal way for animals to move over land. To us the smooth gliding of snakes, without legs, may seem mysterious until we realize what is happening. Seeing a snake for the first time, you might almost imagine that tiny wheels must be

In serpentine crawling, waves pass backward along the body to move the animal forward. The snake in the photograph is a death adder (*Acanthophis antarcticus*).

hidden under its belly, but the actual mechanisms need neither legs nor wheels. In the most usual method of crawling used by snakes, the body forms an S, or a more complicated wavy shape, and then slides forward along the wavy path of its own curves. The head slides forward forming new waves, and the tail slides forward losing waves from the rear. It is as if a wavy line were drawn on the ground and then the animal slid forward along it, every part of the body following all the curves of the line.

The mechanism is simple. The snake forms its body into waves, winding between stones, tussocks of grass, or even slight irregularities on the ground. It then makes the waves travel backward along its body, from head to tail. The body can slide past stones or other obstacles more easily than it can rise over them, so the waves remain stationary on the ground and the snake moves forward. This method of crawling works well on rough ground but not on smooth surfaces

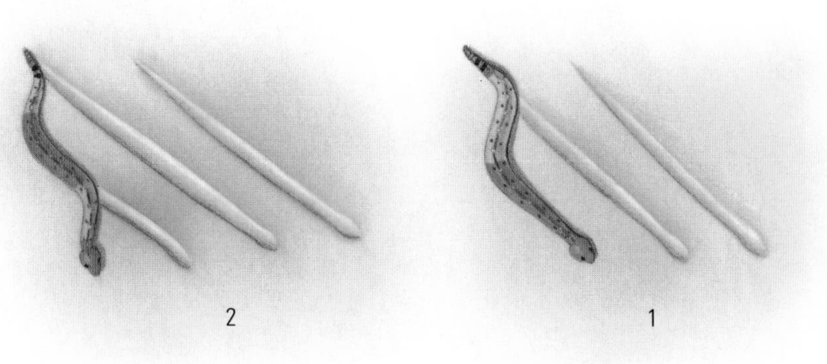


Sidewinding is typical of rattlesnakes, but is also used by some other snakes to travel fast over loose sand. The snake in the photograph is Peringuey's viper *(Bitis peringueyi)* in the Namib desert. Marks in the sand show where its body touched the ground. In the diagram, the tinted parts of the body are on the ground.

that have no obstacles for the snake to push against. A snake on a slippery floor gets nowhere.

This same mechanism is also effective for burrowing through loose sand: less force is needed to push the snout forward into the sand than to push all the waves of the body backward through the sand. The movements are like those of a swimming eel, and this method of burrowing is sometimes described as swimming through sand.

Only a few snakes burrow, but it seems likely that snakes evolved from burrowing lizards. Many lizards live in sandy deserts, where they are in danger of overheating in the sun. They can escape this danger by burrowing in the heat of the day and coming to the surface only when it is cooler. Even quite shallow burrowing is enough to make a big difference in temperature. For example, 2.5 centimeters below the surface of bare soil in a hot part of Australia, the temperature fluctuates at midsummer between a daily minimum of about 25°C and a maximum of 53°C, but at a depth of 30 centimeters in the same soil the fluctuation is only between 35 and 38°C.

A lizard with normal-sized legs could fold them against its sides and swim through sand, but they would be rather in the way. Burrowing is easier without such appendages, and probably for this reason many desert lizards have reduced legs or no legs at all. The legless ones are easily mistaken for snakes, but there is more to being a snake than just having no legs. The easiest way to tell a lizard from a snake is to look at the scales along the underside. A lizard has irregular scales there, but a snake has a single row of large rectangular scales.

The wavy motion of snakes and legless lizards may remind you of the side-to-side bending movements that legged lizards make as they run, but there is a fundamental difference. The legless animals form traveling waves: each wave crest starts at the head and travels back along the whole length of the animal. The legged animals, however, form standing waves: crests form at one or two particular points on the body and do not move. Standing waves are fine for running but would be ineffective for burrowing.

The traveling-wave style of crawling that has been described so far is the most usual gait of snakes and is known appropriately as serpentine crawling. It works well on firm, rough ground, but less well on loose sand, which offers no fixed points for pushing on, and this method would be no use at all for climbing vertical tree trunks.

The successive positions of a standing wave *(left)* and a traveling wave *(right)*.

For travel on these surfaces, two other snake gaits have evolved, both also formed by traveling waves. Sidewinding is the gait that rattlesnakes use to travel over loose sand. Bends travel backward along the body as in serpentine crawling, but there is no sliding of the belly over the ground. Instead, each part of the body is stationary while on the ground but forms new curves in the air as it is lifted periodically to a new position. Marks left in the sand behind the snake show where the body has lain.

A quite different technique (again using bends) enables snakes to climb up grooves in the trunks of trees or fissures in rock. Parts of the

Concertina locomotion enables snakes to travel along narrow crevices. The bull snake (*Pituophis* species) in the photograph is using the technique to climb a tree, jamming itself between the ridges in the bark.

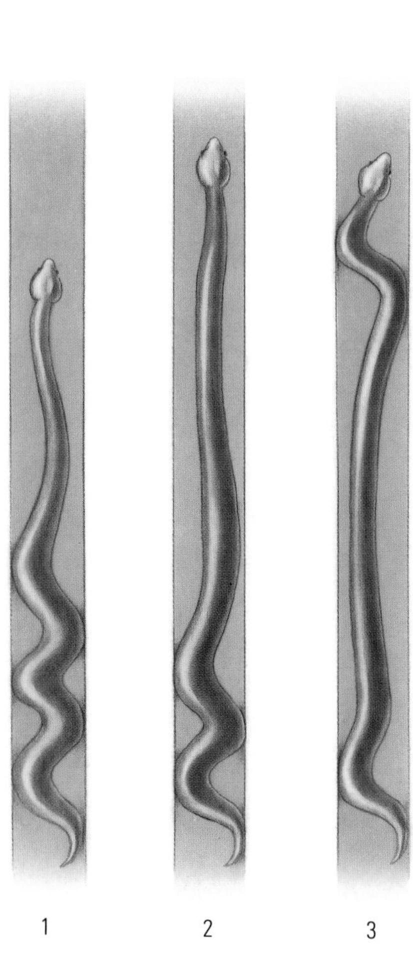

body are folded up like the pleats of an accordian to wedge them tightly in the groove or fissure, an arrangement that has inspired the name "concertina locomotion" for this style of travel. At the front of a wedged region the folds are opening out, pushing the snake's head forward. At the back of a wedged region new folds are forming, drawing the tail forward. Thus groups of folds seem to travel backward along the length of the snake.

More energy is needed to drag a box along a road than to pull a wheeled cart of the same weight, because the sliding of the box is resisted by friction. Using similar reasoning, you might assume that the sliding motion of a snake would require more energy than walking at the same speed, but you would be wrong. Michael Walton, Bruce Jayne, and Al Bennett at the University of California, Irvine, have measured the rates of oxygen consumption of black racer snakes crawling on a moving belt. When the snakes traveled by serpentine crawling, the team of scientists found that the energy cost per unit distance was about the same for the snakes as for lizards and mammals of equal mass (100 grams). The snakes were slow (they could manage short bursts at 1.5 meters per second, but the highest speed they could sustain was only one tenth of that), but they were not uneconomical. Concertina locomotion was much slower and also used more energy per unit distance.

## Stretching and Squeezing

All the movements that we have discussed so far have been powered by muscles that pull on jointed skeletons. Our legs and those of other mammals and of insects have skeletons of stiff rods or tubes, jointed together. Even backbones (including those of snakes) consist of rigid vertebrae connected by movable joints. Without our skeletons we would be flabby and ineffective, but worms, and molluscs such as slugs, move very effectively without any such stiffening.

Earthworm burrowing works on the same principle as the concertina locomotion of snakes: some parts of the body are jammed tightly in the burrow while others move forward. The earthworm, however, is jammed into the available space by making the body swell, not by throwing it into folds. An earthworm's body consists of a line of about 150 segments, which appear as rings on the outside of the body.

Inside, the fluid-filled body cavity is divided into more or less watertight compartments, one for each segment. The body wall includes two layers of muscle, one of longitudinal fibers running lengthwise along the body and one of circular fibers running circumferentially. When the longitudinal muscle of a segment contracts, that segment gets shorter, but because its fluid contents cannot escape it also gets fatter. When the circular muscle contracts, the segment gets thinner but also longer. The segment becomes short and fat or long and thin as the two sets of muscle contract in turn, but its volume remains constant.

In the diagram on this page, segments 7 to 16 and 24 to 29 are short and fat, jammed in the burrow. Segment 7 is about to lengthen and push segments 1 to 6 forward, driving the worm's head onward through the soil. Segment 17 is about to shorten and jam itself tightly in the burrow, and as it shortens it will pull the segments behind it forward. Thus the segments that are fat at any particular instant are stationary and those that are thin are moving forward. Each segment moves forward intermittently.

For this method of burrowing to work as described, the segments at the front of a fat region must always be getting thinner, becoming part of the thin region in front, and the segments at the front of a thin region must be getting fatter, joining the fat region in front. Thus waves of thickening must travel backward along the body.

Earthworms spend most of their time underground, but the movements that serve for burrowing also enable them to crawl on the

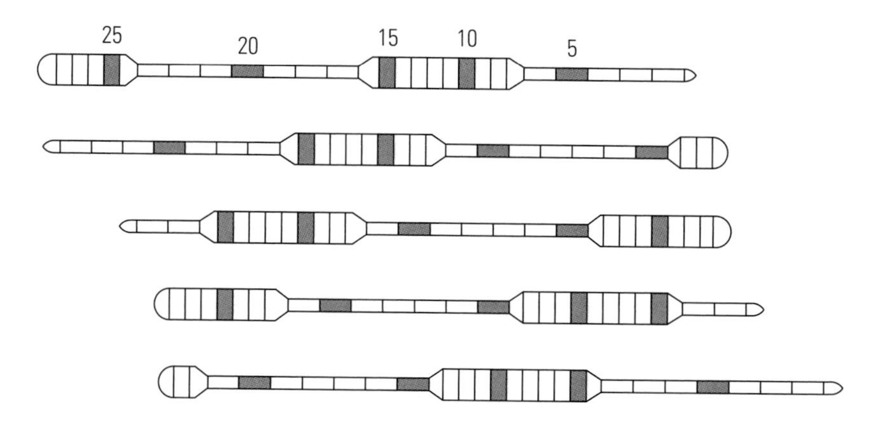


Earthworms crawl by passing waves of muscular contraction backward along the body.

surface. The reason is partly that the thick parts of the body rest on the ground and are held in place by friction, while the thin parts are raised or rest more lightly on the ground. In addition, the worm is prevented from slipping backward by bristles that protrude slightly from the underside of the body, tilted at an angle that lets the worm slide more easily forward than backward. You can feel these bristles if you stroke the underside of a worm with your fingertip. The skin feels relatively smooth as your finger moves backward, but rougher as it moves forward, catching on the bristles. These bristles work like the scales on cross-country skis, which allow the skis to slide forward freely but prevent backward sliding as you climb a slope. The same principle is applied by ratchets that enable machinery to rotate in one direction but not the other.

A garden snail *(Helix aspersa)* on a windowpane. The waves of muscle activity can be seen as faint stripes across the sole of the foot, and the mucus trail is visible behind the snail.

The crawling of slugs and snails looks even more mysterious than that of snakes and worms, for the animals glide forward without obvious movement of the body. However, if you allow them to crawl on a sheet of glass and you watch them from below, you will see something happening on the undersurface of their bodies, or the sole of the foot (as it is called even though there is no leg). A pattern of darker and lighter bands moves along the sole as the animal crawls. Other gastropod molluscs such as limpets, whelks, and periwinkles also crawl on their feet and show similar patterns of bands, with some variations. In limpets the bands are split into left and right halves: bands having a dark side on the left and a light side on the right alternate with bands having the sides reversed. As the animal crawls forward, these bands travel backward along the foot. In slugs each band is light or dark across the whole width of the foot and moves forward. The bands move over the sole of the foot at a rate considerably faster than the mollusc travels. For example, the bands on the foot of a slug crawling at a typical speed of 2 millimeters per second move forward over the foot at 7 millimeters per second. Other molluscs show slightly different patterns of bands, but whether the bands move backward or forward, the effect is always to move the animal forward.

The moving bands suggest the possibility that limpets may crawl essentially as earthworms do. The bands would be waves of muscular contraction formed by alternate bands shortening and elongating. The shortening bands might swell, lifting the elongated ones off the ground. The lengthening of a band would push on the elongated band ahead of it and, unattached to the ground, the elongated band would move forward easily. A shortening band would pull the elongated band behind it forward. The same reasoning as we used for earthworms tells us that if the contractions travel backward along the foot, the limpet will move forward.

A simple experiment seemed to confirm that limpets do move by muscle contractions in the same manner as earthworms. Marks were made on the feet of limpets, and the limpets were then filmed from below as they crawled on glass. Each mark remained stationary while its part of the foot was short and presumably resting on the ground, and moved forward while its part of the foot was elongated. Similarly, in crawling earthworms the contracted (fat) segments are anchored and the elongated (thin) ones move forward.

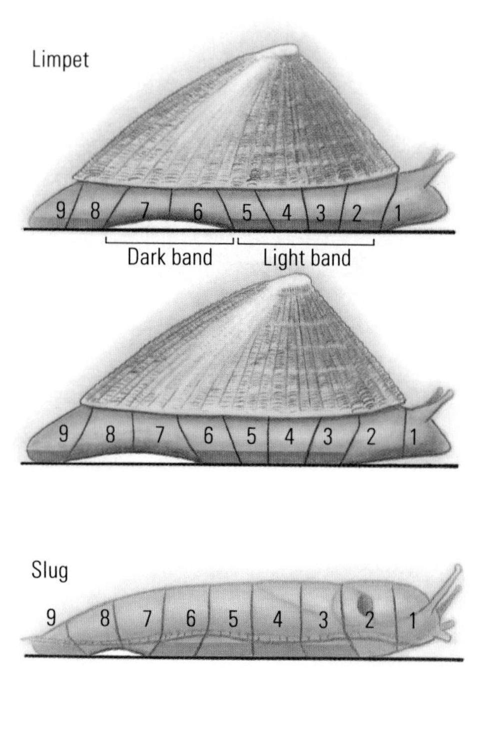


Limpets and slugs were thought to lift parts of the sole of the foot off the ground as they crawled. The elevated sections would be pushed or pulled by waves of contracting muscles.

The theory seemed good for limpets, but how about for slugs, which crawl by waves that move *forward* along the foot? Films of slugs and snails with marks on their feet showed that the marks remained stationary while their part of the foot was elongated and moved forward while it was contracted—just the opposite of what happens in limpets.

Perhaps some of the puzzle could be resolved by taking a closer look at the sole of the foot through a microscope. To catch the sole in midmotion, the slug would have to be frozen in the act of crawling. A good specimen was obtained by picking up a crawling slug and dropping it into a container of liquid nitrogen, where it froze immediately. It was allowed to thaw in a solution of a chemical fixative that coagulated its proteins, preserving the shape of the soft tissues. Part of the slug was then sliced into thin sections, and the sections were examined under a microscope. They showed that the surface of the foot had ridges running across it, as suggested in the theory of limpet crawling. However, the tissue of the projecting ridges that presumably rested on the ground was elongated and that of the furrows contracted; again, this pattern is the opposite of the supposed pattern in limpets.

A little thought shows that the crawling of slugs could be explained by a mechanism that is only a little different from that of limpets. Suppose that the extended bands of the foot are anchored by resting on the ground. Suppose also that the rear edge of each anchored band is shortening and that the rear edge of each raised part is lengthening. The raised parts will be moved forward, but the waves will also travel forward because it is at its rear edge that each band is turning into the other kind.

This explanation of mollusc crawling seemed to be further confirmed by the arrangement of muscle fibers within the foot. There are no distinct layers of fibers running in different directions, as in earthworms. Instead, the muscular foot consists of interwoven fibers running in various directions. In limpets, some of the fibers run vertically through the thickness of the foot and others transversely across its width. If the vertical fibers contract in part of the foot, squeezing it thinner, that part must grow longer or wider or both in order to retain constant volume. If at the same time the transverse fibers prevent that part of the foot from widening, it has no other alternative but to become longer. Thus we can expect contraction of the vertical fibers to lengthen part of the foot, and because the fibers pull upward they will lift it off the ground. The muscle fibers of slugs have a different arrangement; instead of running vertically and horizontally, they slope, some pulling obliquely up and forward and others up and back. When these muscles contract, they *shorten* their part of the foot and lift it off the ground. Each mollusc thus has the arrangement of muscles that seems to be needed for its style of crawling.

The observations of the animals' movements, of the waviness of the feet when they were quickly frozen, and of the arrangement of muscles seemed to add up to a clear, coherent story that told how molluscs crawl. The explanation seemed complete until the work of an American graduate student forced some drastic rethinking.

While still an undergraduate at Duke University, Mark Denny had made his name as a research worker through a beautiful study of the properties of spider silk. He went on to work for his doctorate in Vancouver, and his work on slugs there, published in the early 1980s, gave us a new understanding of mollusc crawling.

Denny saw two problems with the theory. First, it failed to explain the trail of slime (mucus) on which slugs and snails crawl. That slime consists of water and dissolved salts, with a small but significant proportion (3 to 4 percent) of glycoprotein, a compound of protein with sugar molecules. (The mucus that runs from your nose when you have a bad cold has a similar composition.) The mucus presumably did something useful, since the animal would save water and protein by not producing it, but the theory did not explain its function.

The second problem that Denny saw with the theory was that huge forces would be needed to lift the foot from the surface whenever a muscular wave passed. If you lay damp sheets of glass on top of each other, they are very hard to separate (except by sliding) because a thin layer of water forms between the sheets that glues them together very effectively. If the sheets were smeared with mucus instead of water, separating them would be even more difficult, because mucus is viscous like molasses. The flexible feet of molluscs fit very closely onto rocks and other hard surfaces, and the short distance between the two surfaces ensures that the molluscs are firmly glued down. Some limpets of 3 centimeters diameter can hold on to rocks against forces of more than 200 newtons (45 pounds force).

Denny performed a very simple experiment to check whether the foot really is lifted as the muscular waves move along it. He persuaded a slug to crawl on metal foil and then froze it suddenly without detaching the foil. Only after the specimen had been chemically fixed and thawed was the foil peeled off. When the sole of the foot was sectioned and examined microscopically after this treatment, it was found to be flat. There were lengthened and shortened bands running across it, but they were not raised into ridges. Denny concluded that there are no ridges on the feet of crawling molluscs and that the ridges seen previously had formed only after the animal was pulled off the surface on which it had been crawling.

If the whole foot lies flat on the ground, why do some parts move forward while others remain anchored? There is no obvious ratchet like the bristles of earthworms. Denny looked for an answer by investigating the mechanical properties of mucus.

To understand his experiment we must be clear about the difference between elasticity and viscosity. Let us imagine a simple experiment. A layer of rubber is sandwiched between two steel plates and glued firmly to both. After fixing the lower plate rigidly to a table top, you push the upper one horizontally, distorting the rubber layer. The farther you push it, the bigger the force you feel on your finger. If you hold the plate in position the force remains, and if you release it the rubber springs back to its original shape, returning the steel plate to its original position. This is elastic behavior.

Now imagine the same experiment with a layer of molasses (treacle, to British readers) between the plates instead of rubber. A small force will move the upper plate slowly and a large force will move it faster, but if it is kept moving at constant speed the force remains constant. If you stop moving the plate, in any position, the force disappears, and the plate does not spring back to its original position when released. This is viscous behavior. The force does not depend on how far the plate has moved, but on how fast the plate is moving.

Denny collected a supply of slug mucus and performed essentially the same experiment on it. He put the mucus between two metal plates, one flat and one slightly conical, that formed part of an instrument known as a cone-and-plate viscometer. The flat plate was rotated, and the torque needed to prevent the cone from rotating was measured. In rotating the plate, Denny wanted to simulate the conditions under the foot of a crawling slug, where each point on the foot moves at one stage of each muscular wave and remains stationary at another. Therefore, he rotated the plate at constant speed for one second, then held it stationary for a second, then rotated it for another second, and so on.

If the experiment had been performed with an elastic solid such as rubber instead of mucus, the torque would have increased when-

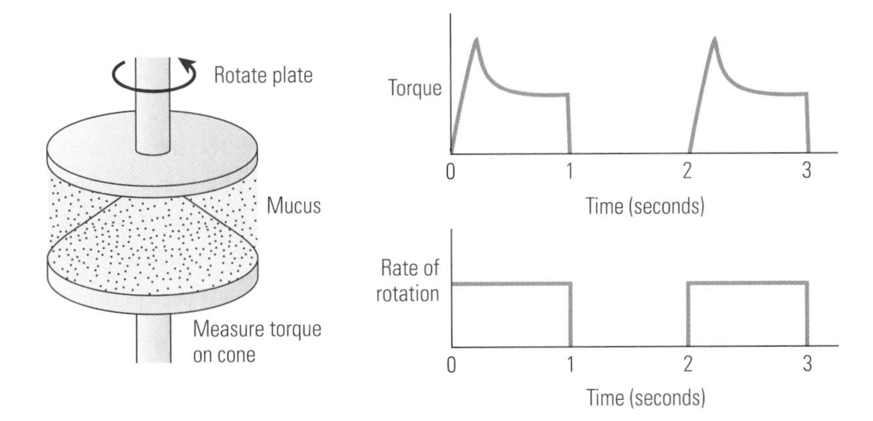


Denny studied the properties of mucus by testing it in a viscometer *(left)*. He found that during steady rotation, the torque rose sharply at first, as if mucus were an elastic solid, then fell somewhat and remained constant, as if it had been transformed into a viscous liquid *(right)*.

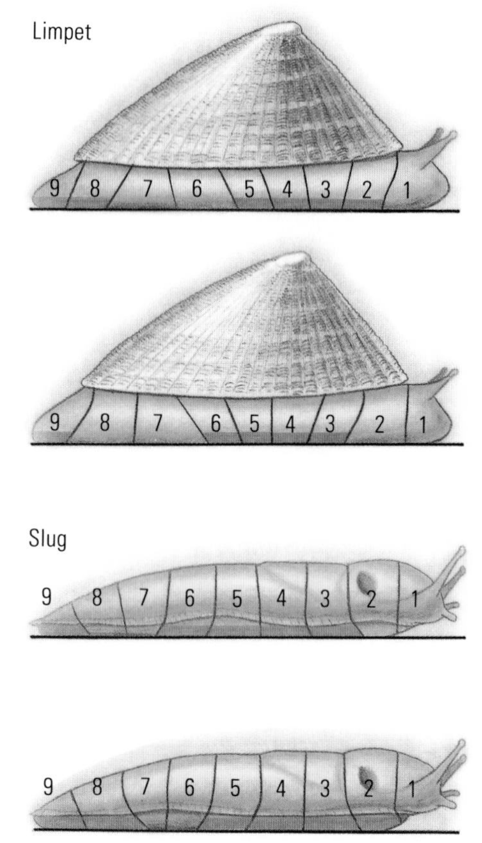


How limpets and slugs crawl according to Denny's theory. Parts of the foot (color shading) are kept anchored by solid mucus underneath, while others are able to slide forward because the mucus under them has become liquid.

ever the cone was rotating and would have remained constant while it was stationary. If the experiment had been performed on a viscous liquid such as molasses, there would have been a constant torque whenever the cone was turning, and that torque would have fallen to zero whenever the cone stopped. The actual result resembled neither of these possibilities. In the early stages of each rotation, the torque rose progressively as if the mucus were an elastic solid. After a while, however, it fell a bit and then remained constant for as long as rotation continued, as if the mucus were a viscous liquid. The period of rest between rotations was apparently enough for the mucus to revert to its original elastic state.

Denny suggested that the state of the mucus under a slug's foot changes with the passage of each muscular wave. Under the extended parts of the foot (which remain stationary), it behaves like an elastic solid, but under the contracted, moving parts it behaves like a viscous liquid. This switch in properties might be expected to happen because the total area of extended parts, at any instant, is greater than the area of the contracted parts. The force needed to move the contracted parts forward must be balanced by the anchoring force on the extended parts (otherwise the extended parts would slide back). The same force must act on both, and so if the contracted parts have less area, more stress (force per unit area) will act on them. The stress under the contracted parts will build up to the level needed to liquefy the mucus, while the mucus under the extended parts remains solid. As the waves travel along the foot, there will always be solid mucus under the extended parts, anchoring them, and liquid mucus under the contracted parts, allowing them to slide forward. The changing properties of the mucus have the same effect as would a ratchet under the foot.

The same mechanism can also work for the backward-moving waves of limpets, if the total area of the parts of the foot that are extended at any instant is less than the area of the parts that are contracted. In that case, the contracted parts will be anchored and the animal will move (as it does) in the direction opposite to the waves.

The molluscs' manner of crawling seems to have severe disadvantages. A snail's pace is proverbially slow, 2.5 millimeters per second (one tenth of an inch per second) or less, a hundred times slower than some beetles. Snails are inevitably slow, because if they were moving fast very large forces would be needed to overcome the viscosity of the mucus under the moving parts of the foot. A second disadvantage is that the method of crawling is expensive of energy and materials. Measurements of oxygen consumption show that the metabolic energy cost per unit distance is about 12 times higher for a slug crawling than for a mouse of the same mass running. For every 10 meters that it travels, a 15-gram slug loses about 1 gram of water in its mucus and 30 milligrams (0.03 gram) of glycoprotein.

Costly though it may be, slug and snail crawling has one very clear advantage. The animal is quite firmly stuck to the surface it is crawling on, at all times. Limpets are not easily dislodged by waves, and a snail can climb up a plant stem without much danger of falling off.

In this chapter, we have learned about many strange animals. They have included fleas and locusts with built-in catapults; gibbons that pump their legs like children on swings; squirrels that can turn their feet back to front; lizards that walk on the ceiling; and slugs that make their slime change from solid to liquid and back again. There are stranger things still in the chapters to come.

# 4

# Gliding

Flying squirrels, which lack the refined aerodynamic design of birds, use gliding as a fast and energy-saving means of travel through the forest. This is *Glaucomys volans,* photographed in Ohio.

We humans tend to think of walking as the normal way for animals to travel. To us, flight seems a remarkable accomplishment, perhaps because we ourselves cannot fly without artificial aids such as airplanes. However, flight is a normal means of travel for a very large proportion of animal species. Most insects fly, and there are more species of insects than of all other animals put together. Most birds fly, and there are twice as many species of birds as of mammals. Even within the mammals, 23 percent of species (the bats) are excellent fliers.

Most flying animals depend almost entirely on powered flight, propelling themselves by flapping their wings. Many large birds, however, are able to travel through the air while holding their wings still. These birds often soar, using natural air movements to keep themselves airborne. Vultures and albatrosses are the largest and most spectacular soaring birds, but gulls and crows also are often seen soaring.

Though flapping is the common style of flight, we will discuss gliding and soaring first, because their simplicity makes them much easier to explain. Gliding and soaring depend on the same aerodynamic principles as airplanes, and these principles are well understood. Flapping introduces formidable complications, as we will see in the next chapter.

## The Aerodynamics of Flight

A bird or an aircraft in flight is held aloft by forces created by the flow of air over the wing. It would be awkward, if not impossible, to measure the forces on the wings and the flow of air around them

This soaring Rüppell's griffon vulture *(Gyps ruepellii)* is an African species with a wing span of about 2.4 meters (8 feet).

while the flier was in midair; instead, these forces are often investigated in a wind tunnel. In flight, the wing is generally moving and the air may be still, but the same force would act on the wing if it were stationary and the air moving, provided the difference in velocity between the wing and the air were the same. Measurements on a wing held stationary in a steady stream of air inside a wind tunnel will reveal the same forces that permit a human-made glider to soar over a hillside or a seabird to glide elegantly over the waves.

Wings are designed to give lift, the force that supports the weight of birds and aircraft in flight. Lift is obtained by pushing air downward: if the wing pushes downward on the air, the air pushes upward on it. Air approaching the wing horizontally is deflected, so that it

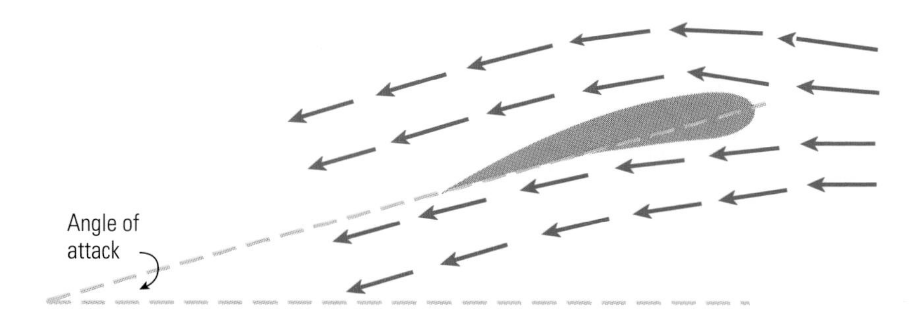


A wing (here seen in section) set at an angle of attack to oncoming air deflects the air downward, gaining upward lift.

leaves the wing on a downward-sloping path. The wings both of aircraft and of birds have an asymmetrical shape, flatter below and more convex on top, that encourages the air to flow downward, but the main reason the air is deflected is that the wings are tilted at an "angle of attack" to the oncoming air. The greater the angle of attack, within limits, the more the air is deflected downward and the greater the lift, but if the angle of attack is increased too far, the aircraft or bird will stall. The flow breaks away from the upper surface of the wings and is deflected downward less effectively, and the lift falls abruptly.

For any particular angle of attack, faster flow gives more lift. Lift is approximately proportional to the square of speed, so twice the speed gives four times the lift. At any particular speed, the maximum possible lift is obtained when the angle of attack is very slightly less than the angle at which stalling would occur. This maximum lift is proportional not only to the square of the speed, but also (for similar wings of different sizes) to the area of the wings:

$$\text{Maximum possible lift is proportional to (wing area)} \times \text{(speed)}^2$$

When an aircraft or a bird flies horizontally, the lift must balance its weight. Instead of asking how much lift an aircraft or a bird can produce at a particular speed, we can ask how slowly it can fly and still get enough lift to support its weight. Since weight and lift must balance, the statement above implies that

Weight
is proportional to
(wing area) × (minimum possible speed)$^2$

Minimum possible speed
is proportional to
$\sqrt{\text{weight/(wing area)}}$

The quantity under the square root sign (the weight that has to be supported, divided by the area of the wings) is called the wing loading. The bigger it is, the smaller the wing area in proportion to body weight and the faster the aircraft or bird must fly. Large wings allow aircraft or birds to stay aloft at lower speeds.

The lift on a wing acts at right angles to the direction of motion through the air. Another component of force, called drag, acts backward along the direction of motion, and it is against drag that the engines of an airplane have to work. When an aircraft is flying horizontally, its weight is balanced by the lift on the wings, and the drag on the wings and fuselage is balanced by the thrust of the engines. Lift is generally useful (it keeps the aircraft in the air), and drag is

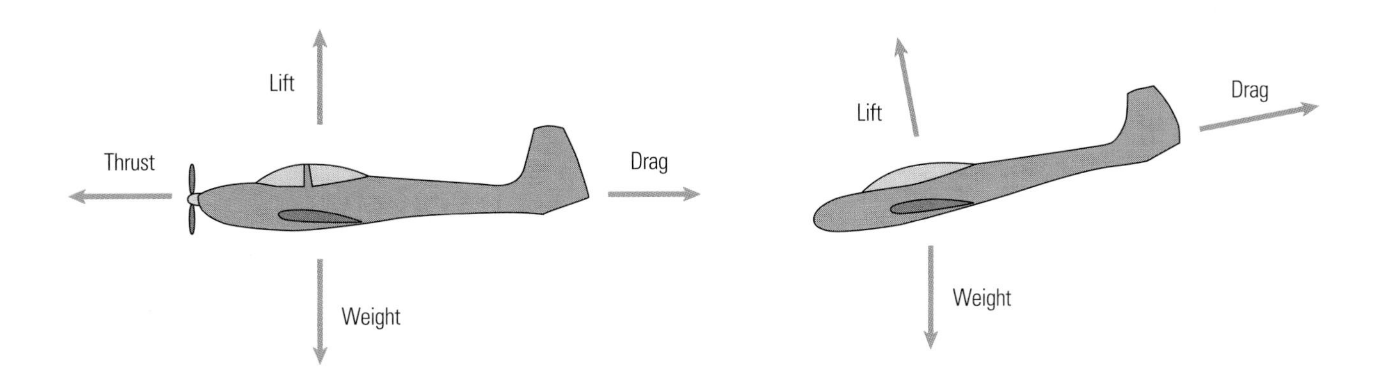


Forces on a powered aircraft *(left)* and a glider *(right)*.

generally unwelcome (the engines have to work against it). Wings are therefore designed to give as much lift, and as little drag, as possible.

Increasing the angle of attack increases the drag as well as the lift, and the ratio of lift to drag is highest at moderate angles of attack. The advantage for an aircraft or a bird of having a high ratio of lift to drag is that it can stay in the air for a lower energy cost. If the wing is well designed, the lift may be as much as 20 times the drag, for wings at least the size of bird wings moving at appropriate speeds. (The wings of small insects give lower ratios of lift to drag.)

Aerodynamicists think of the drag on an aircraft as being the sum of two parts: a profile drag that would act even if no lift were being produced, and an induced drag that acts because the airflow is being deflected to obtain lift. As air encounters the wings, the air is set moving at right angles to the path of the wings, and setting it moving involves giving it kinetic energy. The work needed to give it that kinetic energy must come from somewhere: it is supplied by the engine working against induced drag. The power output of the engines can be divided into profile power, needed to overcome profile drag, and induced power, needed to overcome induced drag.

An aircraft could get the lift it needed by giving a lot of air a small downward velocity or a little air a large downward velocity. An aircraft flying fast passes through more air in each second than if it flew slowly. It thus gives the air less kinetic energy for the same lift, making the induced power less. An aircraft flying slowly has less air available to push on, so it has to give the air a higher downward velocity and needs more induced power. Induced power is therefore high at low speeds and decreases with increasing speed. The reverse is true of profile power, the power that would be needed to drive the aircraft through the air even if no lift were needed. Profile power increases with increasing speed.

The total (induced plus profile) power needed to propel an aircraft is high at low speeds, when induced power is high, and is also high at high speeds, when profile power is high. At an intermediate "minimum power" speed, the total power requirement is least. That is the speed at which a pilot should fly if the aim is to stay airborne for as *long* as possible without running out of fuel. If, however, the aim is to travel as *far* as possible on a tank of fuel, it is better to travel a little faster, at the "maximum range" speed. A third important speed for

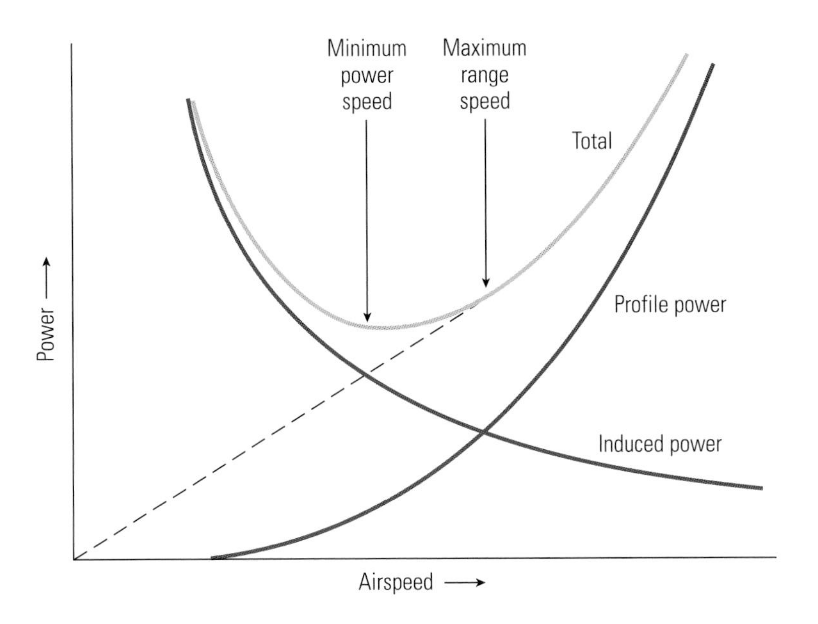


The power needed to propel an airplane is the sum of profile power (which increases with speed) and induced power (which decreases). The total power is least at the minimum power speed. The energy needed to travel a unit distance equals power/speed (because power is energy/time and speed is distance/time) and is least at the maximum range speed. The ratio power/speed at any point on the total power curve is the slope of a line to that point from the point of zero speed and zero power. The broken diagonal line is the least steep such line, so it indicates the maximum range speed.

aircraft is the stalling speed, the lowest speed at which flight is possible. We have already seen that the stalling speed is proportional to the square root of wing loading, and the same is true of the minimum power speed and the maximum range speed. Large wings (giving a low wing loading) enable aircraft to fly slowly and to do so economically. Small wings (giving a high wing loading) enable them to fly economically at high speeds.

Level flight at constant speed requires a balance of four forces: weight, lift, drag, and thrust. In gliding there is no thrust, so the remaining three forces must balance each other. A glider in still air travels on a downward slope, so lift and drag are tilted: lift acts forward and up, and drag acts backward and up. Together they balance the glider's weight. As the glider descends it loses potential energy (energy of height), and the lost energy supplies the work that has to be done against drag.

When a glider glides slowly, it loses height (and potential energy) quickly, because the induced drag is high. When it glides fast, it also loses height quickly, because the profile drag is high. At an intermediate speed, it loses height least quickly and can stay airborne longest:

A glider sinks rapidly in still air if it is traveling slowly or fast, but sinks less rapidly at intermediate speeds.

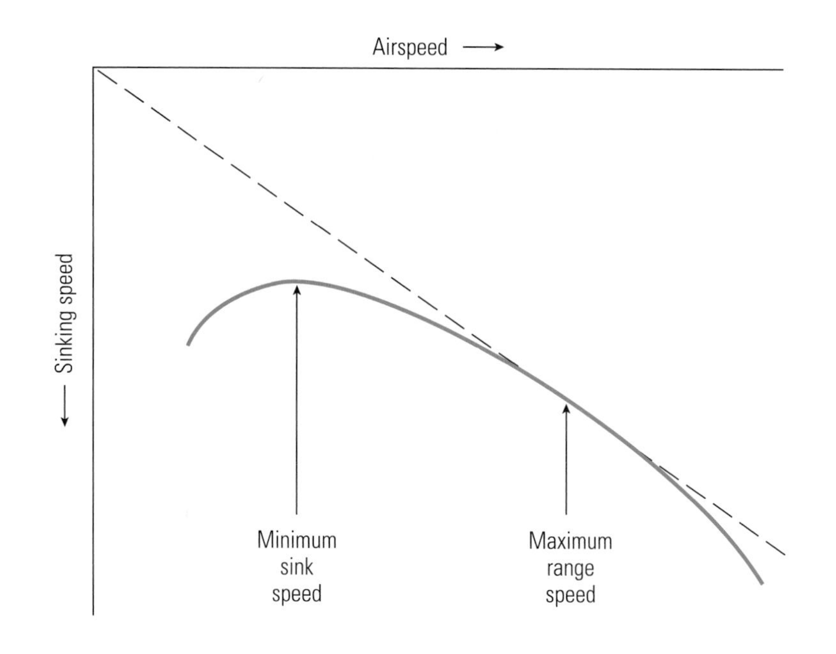


this same speed would be the minimum power speed if the glider were a powered aircraft. To travel the maximum possible distance before hitting the ground, it must glide a little faster, at the speed that would be the maximum range speed for a powered aircraft. As is true for powered aircraft, all the critical speeds (stalling speed, minimum sink speed, and maximum range speed) are proportional to the square root of wing loading. If you want to glide slowly, you will do best with large wings, but to glide fast you need small ones.

## The Gliding Skill of Birds

Much of our understanding of the gliding flight of birds comes from the work of Colin Pennycuick, an English zoologist who is now a professor in Miami. He set out to discover how well birds can glide by comparing their gliding abilities with those of human-made gliders and model aircraft. He started by observing seabirds from a cliff top but soon decided to attempt more accurate measurements in his laboratory, using domestic pigeons. There was not much flying space in his laboratory, so instead of having his birds glide through still air he decided to attempt the flying equivalent of running on a moving belt: he would have a stationary bird gliding in moving air. Instead of the bird moving forward and downward through still air, the air would blow backward and upward past a stationary bird.

Pennycuick built a wind tunnel with a powerful electric fan that drove a jet of air through a nozzle of 1 meter diameter. The nozzle was tapered and grids arranged within to ensure that the air flowed out very smoothly, at almost exactly the same speed everywhere across the width of the nozzle. Now all that Pennycuick needed to perform his experiment was to train a bird to fly in the jet of air, at just the right speed relative to the air to keep it stationary relative to the laboratory. It may seem hard to imagine how a bird could be trained to do that, but Pennycuick achieved it with the help of a teaspoon bowl soldered to the end of a metal tube. He held the teaspoon in the jet of air where he wanted the bird's head to be and rolled chick-peas down the tube so that they landed in the bowl. The only way the bird could get the food was by flying where Pennycuick wanted it to fly.

The wind tunnel was quite large and could blow air very fast, at speeds of up to 20 meters per second (45 miles per hour). Unable to find a suitable room for a tunnel of this size and power, Pennycuick

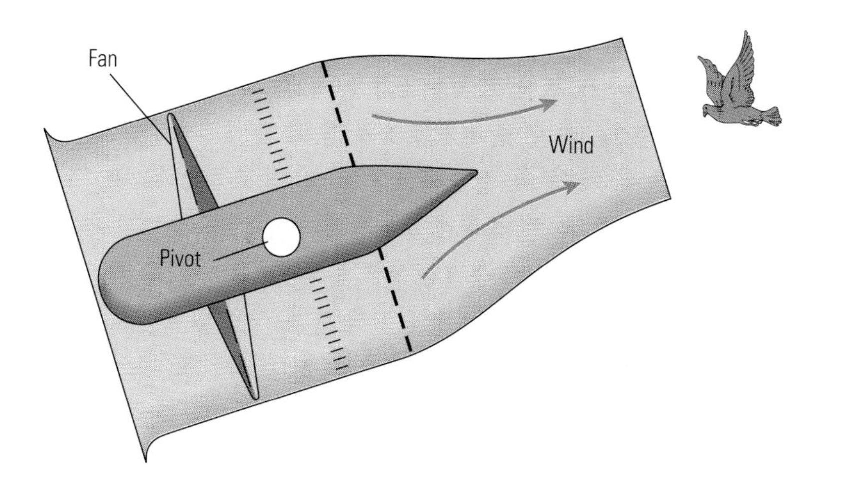


A wind tunnel used for testing the gliding ability of birds.

hung it in the stairwell of a Victorian building at the University of Bristol, where he was working at the time. When he first switched it on it blew out a few windows, but the damage was soon repaired and the experiment worked well.

Pennycuick wanted to test birds as they glided, not as they flapped their wings. If the wind tunnel was tilted up at a sufficiently large angle, which it had been constructed to do easily, and if it was blowing air fast enough, the bird did not need to flap its wings: it could simply glide into the wind (still remaining stationary relative to the laboratory). By varying the speed of the jet and the angle of tilt, Pennycuick was able to find the shallowest angle at which the bird could glide, at each speed. From that he calculated the rate at which it would have lost height if it had been gliding in still air.

The bird could not glide slower than its stalling speed of 8 meters per second (18 miles per hour). At high speeds it lost height rapidly. Like artificial gliders it had a minimum sink speed at which it lost height least quickly, but its performance was poor by engineering standards. Good gliders, both full-size ones and models, lose only

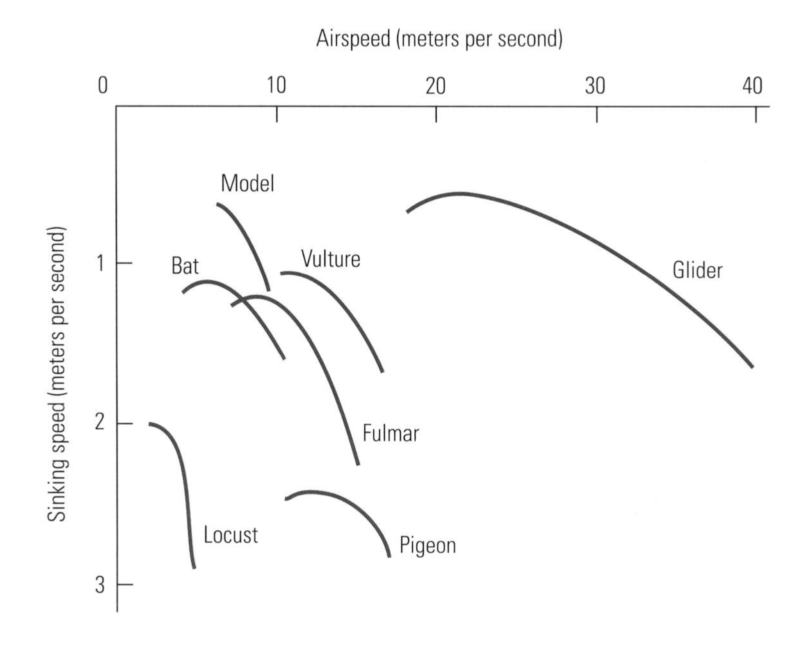


The sinking speeds of gliders, birds, and a fruit bat, gliding at various speeds.

about 0.5 meter of height per second at the minimum sink speed, but the pigeon apparently could lose no less than 2.5 meters per second. It is not clear whether pigeons really are very poor gliders or there was something about the experiment that prevented the pigeon from gliding well.

Since that pioneering study, several similar experiments have been performed on other flying animals. Pennycuick himself moved his wind tunnel to Kenya and repeated the experiment on fruit bats (of about 0.55 meter, or 22 inches, wing span). Vance Tucker of Duke University repeated it with a small species of vulture and with a buzzard. (Chick-peas were no use as an inducement for them.) All these, and fulmar petrels that Pennycuick had observed from a cliff top, did much better than the pigeon but slightly less well than artificial gliders: all lost height at minimum rates of around 1 meter per second. Pennycuick has pointed out that comparisons with gliders are really rather unfair: a gliding bird is not a proper glider, but is more like a powered aircraft with its engine switched off.

Pennycuick is a keen pilot, which is probably why he is so interested in bird flight. He was able to turn his piloting abilities to good advantage during several years he spent in Kenya, where he had a motor glider (an aircraft designed to glide, but with an engine that could be used to get it airborne or to get out of difficulties). Aloft in his aircraft, he glided with the vultures, storks, and pelicans over the East African plains, observing their behavior in the air. By this means he was able to study the ability of these birds to glide for great distances in weather conditions that are especially favorable for that means of travel.

## Thermal Soaring

Although vultures fly for most of the day, they flap their wings very little, keeping themselves airborne by soaring in thermals. These are columns or patches of rising air, formed over ground that has been heated by the sun: the hot ground heats the air, which expands, becomes less dense, and so rises. Thermals are often reasonably easy to find because they form in predictable places (for example, over bare rock). Their tops are often marked by the presence of cumulus clouds (the ones that look like cotton wool). Strong thermals do not appear until the sun is fairly high in the sky, and they disappear in the evening, so thermal soaring is possible only between those times. Vultures spend the night roosting on cliffs or trees.

Vultures and other birds circle upward in thermals to gain the height they need to start a straight glide. They circle by banking, tilting so that one wing tip is higher than the other. The tilt gives the lift a horizontal component that serves as a centripetal force. If the birds tried circling in still air, they would lose height—indeed, they would lose height faster than if they were gliding in a straight path, because circling requires more lift (and therefore more induced drag) than does straight gliding. A bird circling in a thermal is able to rise if

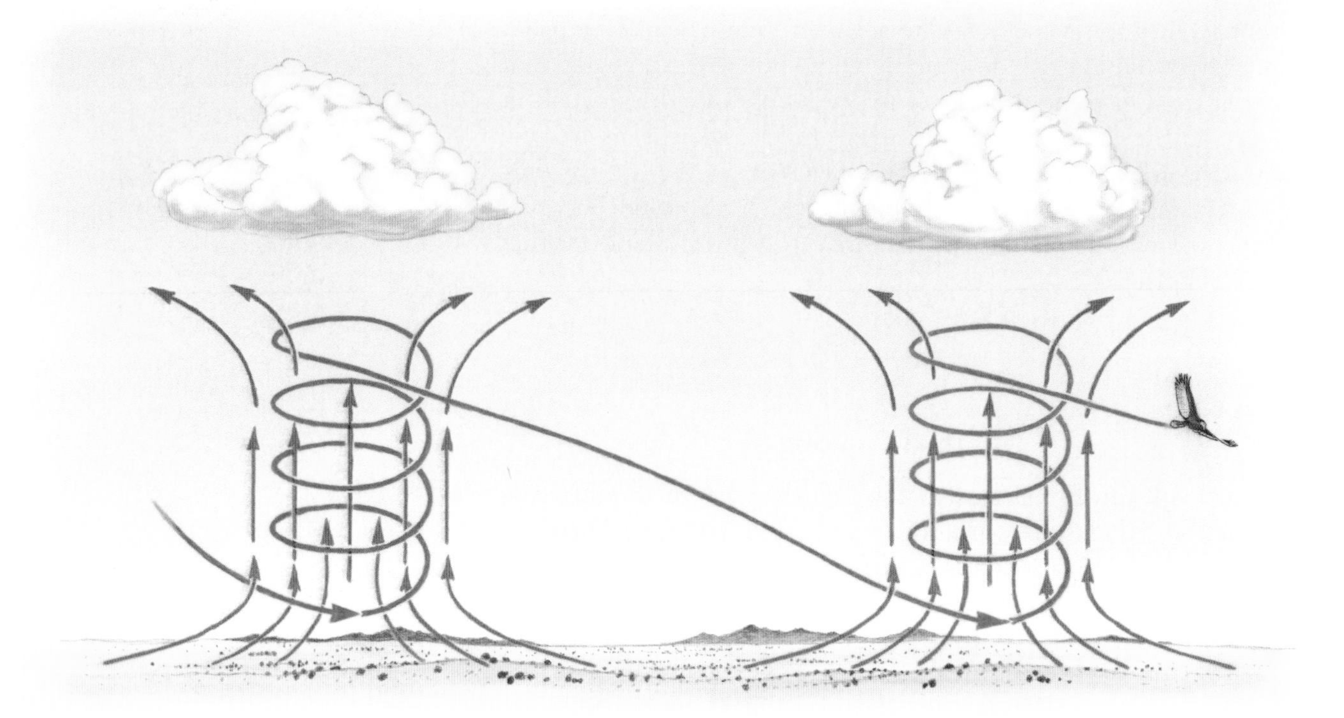

Vultures soar by circling in thermals.

Great white pelicans (*Pelecanus onocrotalus*) soaring in a thermal. This species is widely distributed in Eastern Europe, Asia, and Africa.

the air in the thermal is rising faster than the bird would sink in still air. For example, if a bird in a typical thermal is sinking relative to the air at 1 meter per second, and the air in the thermal is rising at 4 meters per second, the bird will gain height at a rate of 3 meters per second. It is common for soaring birds to gain height at such rates.

Vultures travel by gliding from thermal to thermal, sometimes over long distances. For example, Rüppel's griffon vulture nests on cliffs at the edge of the Serengeti and, in the breeding season, commutes daily between its nest and the herds of wildebeest and other mammals from which it takes its food. The vultures soar over the herds, watching for deaths so that they can feed on the carcasses.

Pennycuick used his glider to follow vultures on their daily journeys. On one occasion he stayed with a Rüppel's griffon vulture for 96 minutes while it traveled back to its nest from the herds. In that time it covered 75 kilometers (47 miles) entirely by soaring, stopping to circle in only six thermals and rising to heights of up to 1500 meters (5000 feet) above the ground.

Even more impressive are the much longer journeys of white storks. These birds glide from thermal to thermal for much of their annual migrations between Europe and southern Africa. A straight route would take them over the Mediterranean, where there are no thermals, but they manage to get help from thermals almost all the way by taking a detour. The western European populations travel over the Straits of Gibraltar and the eastern ones over Suez.

An albatross slope soars along a wave.

A shy albatross (*Diomedea cauta*), so called because it seldom approaches ships. This one is using its feet as air brakes to control its speed.

## Slope Soaring

Slope soaring is an alternative to thermal soaring, much used by seabirds. Gulls can often be seen soaring over a hillside or cliff, gliding backward and forward without beating their wings. They are held aloft by the upward air movements formed as the slope of the ground deflects wind upward. If the air is rising as fast as they would sink in still air, they can maintain their height.

The same technique works well over the open ocean, on the windy side of large waves. Albatrosses soar over the Antarctic Ocean, watching for fish or squid swimming near the surface. When they spot one, they often land on the sea to feed. Although they depend largely on slope soaring, they also use another soaring technique that involves swooping up and down between the strong wind high above the sea and the slower wind near its surface.

Pennycuick has made many observations of slope soaring using an instrument that he invented, called the ornithodolite. It is a combination of theodolite and rangefinder that will measure the compass bearing of a flying bird, its angle of elevation above the horizon, and its distance from the observer: with this information the bird's position can be fixed in three-dimensional space. Pennycuick has only to keep the instrument focused on the bird and press a button whenever he wants to record the bird's current position and the time in his portable computer. In that way he gets the information needed to calculate the speeds of soaring birds and their rates of change of height. He also measures the speed and angle of the wind, using an anemometer (wind gauge) attached to a pole as near as possible to where the birds are flying.

Pennycuick usually encountered the largest of albatrosses, the so-called wandering albatross, soaring between 2 and 12 meters above the waves. The wing span of this species measures more than 3 meters, and these impressive wings hold aloft an adult bird weighing from 8 to 10 kilograms. The bird would typically glide in still air at speeds of about 12 meters per second (27 miles per hour), but when it was slope soaring, gliding into the wind, it traveled faster than 12 meters per second relative to the air and slower than that speed relative to the sea. For example, an albatross gliding into a 10-meter-per-second headwind would typically travel at 16 meters per second relative to the air, gaining ground at $16 - 10 = 6$ meters per second. In very calm conditions, slope soaring could not keep these birds airborne and they had to flap their wings occasionally, but unless it was very calm, they flapped their wings very seldom.

To stay in the upwardly deflected wind over a wave, the bird must soar along the wave—which may not be the direction in which it wants to travel. It can travel in other directions by taking a zigzag course, soaring for a while along a wave and gaining height or speed, then gliding for a while at an angle to the waves before joining another wave and soaring along it. Pennycuick observed albatrosses following a ship in this way and found that the total length of their zigzag paths averaged 1.5 times the straight-line distance that they traveled. On one occasion, for example, his ship was sailing directly into a 6-meter-per-second wind when a wandering albatross overtook it from astern. The bird changed direction every 10 to 15 seconds to follow a zigzag path and achieved a straight-line speed relative to the sea of 6 meters per second, entirely without flapping its wings.

The energy cost of gliding is much less than that of flapping flight, as we will learn in Chapter 5, but even when the wings are not being flapped, metabolic energy is needed to maintain tension in their muscles. The main muscle involved is the pectoralis muscle, which pulls the wings down in the downstroke of flapping flight and holds them in position against the lift forces that act on them in gliding. Vultures, storks, and albatrosses have the pectoralis muscle divided into two distinct parts: a large superficial part that is dark red and a much smaller deep one that is paler. It is believed that the superficial part consists of fibers capable of shortening rapidly to beat the wing and the deep part of slowly contracting fibers that can maintain tension at little energy cost during gliding. Only in soaring birds is the muscle divided into these distinct parts. Albatrosses differ from the others in having what seems to be an even more effective device for saving energy while soaring. A fan-shaped ligament locks the shoulder joint when the wings are horizontal and fully spread. Consequently, the wings cannot be lifted above the horizontal until they are moved slightly back from the fully spread position. If the shoulder lock holds an albatross's wings horizontal, the bird may be able to soar with very little tension in its muscles.

Whereas albatrosses travel long distances by slope soaring, small falcons known as kestrels have a different use for the technique.

An American kestrel (*Falco sparverius*) soaring. Kestrels spend much of their time slope soaring, stationary relative to the ground, watching out for prey.

They hang stationary in the air, ready to pounce on any mice or voles that they see moving on the ground below. To hold themselves in place over level ground they fly into the wind, matching their speed exactly to that of the blowing air. Often, however, they slope soar over sloping ground, keeping themselves stationary without having to flap their wings.

John Videler of Gröningen University in the Netherlands has made many studies of kestrel flight. He showed how precisely they stay in position by filming them with a camera on a tripod, locking the camera rigidly to the tripod as soon as it was focused on the bird. His films show that during 20 or 30 seconds the bird's head often moves less than a centimeter from its initial position relative to the ground. Those observations were of birds flapping their wings against the wind, but Videler has also made careful observations of slope soaring over a sea dike, against onshore winds averaging 9 meters per second. The windward side of the dike sloped at 14 degrees to the horizontal, and at the height where the birds flew (on average, 6.5 meters above the ground), the wind was angled upward at about 7 degrees to the horizontal. When the bird remained stationary in this situation, its movements *relative to the air* and the forces on it were exactly the same as if it had been gliding in still air at 9 meters per second and at an angle of 7 degrees to the horizontal.

Some soaring birds seem to walk on water. These are the storm petrels, small seabirds of 30 to 40 grams (about the mass of a cardinal finch). They hang suspended just above the sea surface with their wings spread but not flapping, dipping their feet into the water from time to time and looking out for the small fish and squid on which they feed. They are using their own peculiar soaring technique, sea-anchor soaring, which works by the same principle as the flight of a kite. Their weight is supported by lift on their wings, while the drag exerted on them by the air is balanced by the forces that the water exerts on their feet. The bird is blown slowly backward over the water, so the drag on the feet is directed forward.

A bird that is good at slope soaring will not be particularly good at thermal soaring, and vice versa, because these two soaring techniques work best at different speeds. Many thermals are only a few tens of meters in diameter, and therefore thermal soaring requires the ability to turn in small circles. If thermal soarers circled at high

The sea-anchor soaring technique of the storm petrel.

speed, they would have to increase the lift on their wings greatly to obtain the necessary centripetal force. As a result, induced drag would increase and the birds would sink faster relative to the air. Accordingly, the best thermal soarers are birds that can circle very slowly. In contrast, good slope soarers can glide fast. To remain stationary as kestrels do, they must be able to glide as fast as the wind. To make headway against the wind as albatrosses do, they must be able to glide faster than the wind. Thermal soarers, then, should be good at slow gliding and slope soarers at fast gliding.

The two different soaring strategies are reflected in the areas of the wings. Both the minimum sink speed and the maximum range speed are proportional to the square root of wing loading, so thermal soarers should have low wing loadings (large wing areas for their weights) and slope soarers should have higher wing loadings. As expected, vultures have much bigger wing areas than albatrosses of the same weight; for example, Rüppel's griffon vultures averaging 7.6 kilograms mass (74 newtons weight) had an average wing area of 0.83 square meter, making the wing loading 90 newtons per square meter. Wandering albatrosses averaging 8.7 kilograms mass had 0.61-square-meter wings, giving a wing loading of 140 newtons per square meter.

Comparisons of this kind must be made between birds of similar size, because different-sized birds of similar habits have different wing loadings. To understand this, imagine two geometrically similar birds, one twice as long as the other. The longer bird will also be twice as wide and twice as high, and so it will be eight times as heavy. Its wings are twice as long and twice as broad, so have four times the area. Four times the wing area has to support eight times the weight; as a consequence, wing loading is twice as high for the longer bird as for the shorter. A smaller marine slope soarer like a 1-kilogram white-chinned petrel might be expected to have half the wing loading of an 8-kilogram albatross, which indeed it has. Similarly, small vultures have lower wing loadings than large ones. For any particular body mass, albatrosses and other slope soarers have smaller wing areas (and so larger wing loadings) than vultures and other thermal soarers.

Wing span is the distance from one wing tip to the other, with the wings spread, and the chord is the distance from the front to the rear edge of the wing. Although vultures' wings have much larger areas than those of albatrosses of equal mass, their spans are a little less. The 7.6-kilogram vultures that we have been discussing had wing spans averaging 2.4 meters, and geometrically similar 8.7-kilogram vultures would have had 2.5-meter spans. In contrast, the 8.7-kilogram albatross had a span of 3.0 meters. Thus vultures have short broad wings and albatrosses have long narrow ones. The difference in shape can be described by calculating the aspect ratio, which is the span divided by the mean chord. The aspect ratio is 7 for the vulture that we have been discussing and 15 for the albatross.

As a general rule, the higher the aspect ratio the better the aerodynamic performance of a wing. The reason is that the bigger the span, the more air is driven downward each second. If two wings of different spans traveling at the same speed have to produce the same lift, the one with the longer span is able to do it by driving a larger mass of air downward, with a smaller downward velocity. It gives less kinetic energy to the air than the wing of smaller span, which gives a larger downward velocity to a smaller mass of air. For this reason, a bird with a longer wing span suffers less induced drag. A glider with a longer wing span will lose height less fast than a similar glider with a shorter span, especially at the low speeds at which induced drag is larger than profile drag.

Simple theory suggests that vultures would fly better if they had longer, possibly narrower wings. However, if, say, an 8.7-kilogram vulture were to be given an albatrosslike aspect ratio without reduc-

A magnificent frigate bird (*Fregata magnificens*), a tropical seabird with very large but narrow wings.

Birds spread their wings more when gliding slowly than when gliding fast. These drawings are based on photographs of pigeons gliding in Pennycuick's wind tunnel.

ing its wing area, it would need wings with a 3.7-meter span, 22 percent more than the span of an albatross of the same mass. Wings that long might be difficult to manage when a vulture leaving a kill was taking off from the ground.

There is one group of birds that combines low, vulturelike wing loadings with high, albatrosslike aspect ratios, but the largest of them have a mass of only about 1.5 kilograms and a wing span of no more than about 2.3 meters, less than some vultures. These are the frigate birds, tropical seabirds with exceptional soaring habits. They live in the latitudes where the trade winds carry cool air from temperate regions over the warm tropical sea. Here, unlike at other latitudes, there are thermals over the sea, created as air is warmed by the sea surface and rises upward. Frigate birds soar in these thermals, apparently remaining airborne day and night, for they travel far out over the oceans but apparently never land on the sea surface. They feed on flying fishes and on squids that leap out of the water, and they also chase other birds and steal food from them. They need low wing loading for thermal soaring, and long wings are probably less of a problem than they would be for vultures, because frigate birds have no need to take off from the ground. They nest in treetops and can gain speed, when they take off, by diving out of the tree.

Not only do different kinds of birds have different wing loadings to suit their flying habits, but by partially folding its wings, an individual bird can increase its wing loading when it wants to go fast. The pigeons that flew in Pennycuick's wind tunnel kept their wings spread to their full 65-centimeter span when gliding at 9 meters per second, but folded their wings progressively as speed increased, until at 22 meters per second the span measured only 25 centimeters. The folding of the wings reduced the wing area from 600 square centimeters with the wings spread at 9 meters per second to 400 square centimeters at 22 meters per second.

## Making a Landing

However fast a bird glides, it must slow down for a safe landing. Birds use their feet as air brakes, holding them close to the body in fast flight but lowering them to lose speed. They cannot reduce their speed too much, however, or they will stall and be unable to produce enough lift to support their weight. There are two features of wing design that seem to enable birds to glide more slowly without stalling than would otherwise be possible. One of these is the alula, a tuft of feathers on the front edge of the wing supported by the bone of a rudimentary index finger. It lies flat against the rest of the wing in fast flight, but at low speeds is lifted and may help to keep the air flowing smoothly over the wing at angles of attack at which stalling would otherwise occur. The same principle is used in aircraft, in the device called a leading-edge slot. The stalling speed is probably also reduced by separating the large feathers at the wing tip, giving a multislotted effect. The feathers at the wing tips of crows, vultures, and many other birds separate in slow flight.

Even birds that have such devices seem to stall deliberately when landing: increasing the angle of attack until the wings stall is a very

Sandhill cranes *(Grus canadensis)* landing in New Mexico.

effective way of increasing drag. Stalling can be detected in photographs of landing birds because the irregular airflow over the upper surface of the wing ruffles the feathers.

There are other ways of slowing down, besides using the feet as air brakes or stalling the wings. Ducks land at high speed on ponds and slow themselves by holding their feet in the water. Guillemots landing on the cliffs where they nest approach at a lower level than the nest and veer upward, slowing themselves by converting their kinetic energy (energy of speed) into potential energy (energy of height) and stalling just before landing.

The birds that we have discussed so far all glide well: they can glide at shallow angles, losing height only slowly. There are other animals that glide much less well but are particularly interesting because they give us hints suggesting how birds and bats may have evolved the power of flight.

## Flying Squirrels

These less good but still very effective gliders include flying squirrels, which look much like ordinary squirrels but have flaps of skin between their fore and hind legs. By spreading their limbs they can open out this skin and stretch it tight, making a surprisingly effective airfoil. Much of what we know of the habits of flying squirrels comes from the observations of Keith Scholey, who while a Bristol University graduate student studied a species that lives in Borneo. His base was a house in a forest clearing, from which he would watch the flying squirrels moving through the surrounding trees. They spent the day in their nests, in holes in dead trees, but came out at dusk to feed on leaves throughout the night. To find young leaves, which they prefer, they often had to travel some distance through the forest, gliding from one tree to the next. (Young leaves can be found on some trees in this tropical forest at almost any time of year.)

As dusk approached each evening, Scholey settled down on a chair on the sun deck, waiting for the flying squirrels to emerge from their holes. With him he had a panoramic photograph of the forest edge and a stopwatch. When the squirrels came out and began to glide, he would mark, on the photograph, the positions on the trees at which each glide started and ended, and he would time the glides with the stopwatch. He watched the squirrels until they disappeared into the forest or it became too dark to see them. On the following day, using surveying instruments, he measured the length of each glide and the height lost.

The flying squirrels traveled by climbing to the top of a tree trunk, gliding to a lower point on a nearby tree, climbing to the top of its trunk, gliding again, and so on. They glided distances of 30 to 130 meters, diving steeply at first to gather speed and then continuing at

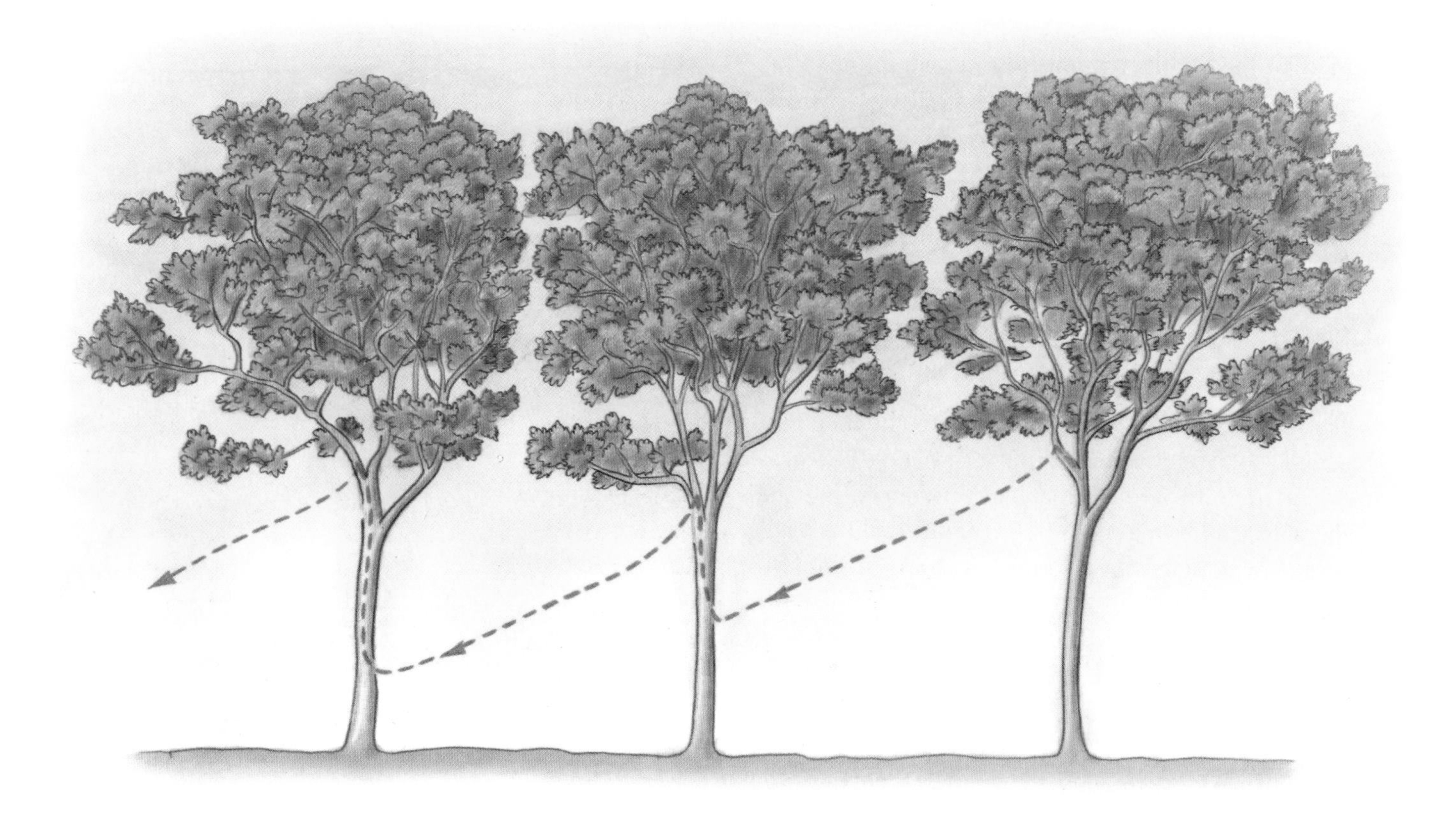

Flying squirrels travel through a forest by climbing the trunks of trees and gliding between trunks.

a shallower angle, about 12 degrees to the horizontal (losing, that means, about a meter of height for every 5 meters that they traveled). They glided fast, at about 15 meters per second, and so lost height at about 3 meters per second. This rate of loss of height is poor in comparison with that of birds, since birds may lose only 1 meter per second when gliding at their minimum sink speed.

The glide has to be fast because wing loading is high. A specimen of this particular species, the giant red flying squirrel, had a body mass of 1.3 kilograms (weight 13 newtons) and a wing area of 0.11 square meter. The wing loading was thus 120 newtons per square meter, high in comparison with birds. (A petrel of similar mass had a wing loading of 80 newtons per square meter, and a buzzard 46 newtons per square meter.)

Scholey estimated that the squirrels lost 6 meters of height in the initial steep dive. A body that starts at rest and falls freely through 6 meters has a final speed of only 11 meters per second. Thus even a dive of this depth is not quite enough to accelerate the animal to 15 meters per second, the speed of most of the glide: the animal must have continued to accelerate in the shallower part of the glide. Because the squirrel needs an initial steep drop to accelerate, it would hardly have been worth gliding if there had not been many trees over 20 meters tall. The flying squirrels' method of travel is much more effective between the tall trunks of a tropical forest than it would be in most North American or European woods.

Near the end of a glide the squirrel is approaching a tree trunk at 15 meters per second and must slow down to avoid a damaging impact. It brakes its onward rush by the same means guillemots use when landing on cliffs, veering upward to convert some of its kinetic energy back into potential energy. The wing stalls as the squirrel slows down, but that does not matter if the animal has timed its action skillfully so that it does not stall until immediately before landing. In principle, the rise at the end of the glide could recover most of the height lost in the initial steep dive, but some height is lost in the fraction of a second before landing, because of the stall.

The squirrels need good control of their flight to reach their intended landing points, steering around any obstacles on the way, and land safely. They display their skill in the breeding season, when they come out of their holes by day and make spectacular glides, apparently as a courtship display. Scholey once saw a squirrel make a midair turn through 180 degrees and land again on the tree from which it had taken off.

Flying squirrels travel reasonably fast. To travel 60 meters, for example, at the gliding speed of 15 meters per second would take only 4 seconds, but a 60-meter glide actually takes 6 seconds because of time lost in the initial dive and in landing. The height lost in a 60-meter glide is about 18 meters (6 meters for the initial dive plus 1 meter for every 5 meters traveled). To recover the lost height the squirrels climb 18 meters up the trunk after landing, which at about 0.7 meter per second takes 26 seconds. Thus the total time needed to travel 60 meters, gliding and then climbing, is about 32 seconds, giving a mean horizontal speed of 1.9 meters per second. It is doubtful whether the squirrel could travel this fast if it did not glide. When an ordinary, nonflying squirrel was trapped and then released and timed as it ran away, it ran at 2.2 meters per second, but that was probably a sprinting speed that could not have been sustained for long. The same species running and jumping from branch to branch through the forest canopy was estimated to travel at only 1.0 meter per second.

## Other Gliders

There are flying lizards in the tropics as well as flying squirrels. *Draco* has simple wings on the sides of its body, consisting of flaps of skin stiffened by extensions of the ribs. Where these ribs emerge from the body, there are joints that enable the animal to spread its wings or to fold them against its sides. Like the flying squirrels, it glides only moderately well, losing 7 meters of height in a 20-meter glide. Because it feeds on common insects, it does not have to commute through the forest to suitable feeding sites, as flying squirrels do. It climbs up a tree, searching the trunk for insects, then glides to the base of the next tree and searches its trunk for food. Gliding enables the lizard to move to the bottom of the next tree quickly, at little energy cost.

Birds and bats must have evolved from ancestors that flew less well than they do, possibly gliding from tree to tree like flying squirrels and flying lizards. It is easy to imagine bats evolving from an ancestor that looked like a flying squirrel and that used its gliding ability either as flying squirrels do, to commute through the forest to trees bearing young leaves or fruit, or as *Draco* does, to get from tree to tree while searching for insects. There are two distinct groups of bats, and they are believed by some zoologists to have evolved separately. The largest number of species (including all the North American and European ones) are microbats. Most of these small bats feed on insects. Microbats usually catch their prey in the air, but they may have evolved from an ancestor with feeding habits like *Draco*'s. The megabats of Africa and Asia are generally larger than microbats, as their name suggests: the biggest have a mass of about 1.4 kilograms and a wing span of 1.2 meters. They eat only plant food (largely fruit) and probably evolved from an ancestor that behaved like a flying squirrel, as well as looked like one.

The flying lizard *Draco volans,* gliding between branches in Borneo. The wings will be folded against the body when it lands.

There are two theories about the evolution of birds. One is that ancestral birds lived in trees, moving from tree to tree by gliding. The other is that they were fast-running animals that lived on the ground. This theory suggests that these ground-dwelling birds leapt from time to time as they ran, spreading their wings and gliding to extend the leap. The problem with that suggestion is that wings were probably small in the early stages of evolution, so the wing loading would be high and the animals would have to travel very fast indeed to be able

*Archaeopteryx*, a bird from the time of the dinosaurs, as it may have looked in life. Some authors believe that early birds lived on the ground, using their wings to aid long leaps as they ran. Others believe that they lived in trees and used their wings to glide between branches.

to glide. For example, it can be calculated from the wing loading of the giant red flying squirrel that it would probably stall at 13 meters per second. No flying bird or small mammal can run at such a speed, which is quite fast even for antelopes.

This chapter has been almost entirely about gliding vertebrates, because few insects make much use of gliding. The reason seems to be that small flying animals have low wing loading, and therefore their minimum sink and maximum range speeds are also low. They glide too slowly to be slope soarers, for they cannot make headway against any but the slowest winds. Although a slow speed is suitable for gliding up thermals, thermal soarers must be able to glide reasonably fast from one thermal to the next. It seems impossible to design a glider that will sink more slowly than 0.5 meter per second in still air, so a slow glider cannot travel very far before hitting the ground and might find it difficult to reach the next thermal. Vultures, storks, and albatrosses are large, and few soaring birds are small.

Despite these impediments, there are a few large butterflies that soar. Monarch butterflies (wing span 11 centimeters, 4.3 inches) soar while migrating between Canada and Mexico, although their

A Monarch butterfly *(Danaus plexippus)* in flight. This species migrates between Canada and Mexico.

maximum range speed is less than 3 meters per second and they lose a meter of height for every 4 meters gliding in still air. If the wind is against them they use flapping flight, keeping close to the ground where the wind speed is low, or else they do not fly at all: if they tried to soar against the wind, they would be blown in the opposite direction. On good days when the wind is behind them, however, they may spend as much as 80 percent of the time soaring, circling in thermals and slope soaring over buildings, letting the wind carry them along.

This chapter has shown how much animals can achieve, simply by gliding and soaring. Vultures soar in the thermals over the African plains, and storks migrate by thermal soaring between Europe and Africa. Slope soaring supports kestrels as they watch for prey and enables albatrosses to remain airborne day and night over the Antarctic Ocean. Even the lesser gliding skills of flying squirrels enable them to travel faster through the forest than if they had to run. Though so much can be done by gliding alone, even more is possible for animals that are capable of powered flight, as we shall see in the next chapter.

# 5

# Flapping Flight

Flamingos in flight in Kenya, where huge flocks of these colorful birds feed from the waters of Lake Nakuru.

Airplanes get lift from their wings to support their weight and get thrust from propellers or jet engines to overcome drag. Flying animals, however, must get thrust as well as support from their wings. In this respect, they are rather like helicopters, which depend on their rotors for both support and thrust. It may help us to understand the flapping flight of birds, bats, and insects if we first think about helicopters.

## Flapping to Stay in One Place

The streamlined blades of a helicopter rotor are tilted at an angle to the plane in which they rotate, so that they meet the air at an angle of attack. The blades drive air downward, and the air exerts lift on them in return. To hover steadily in one place on a windless day, the rotor

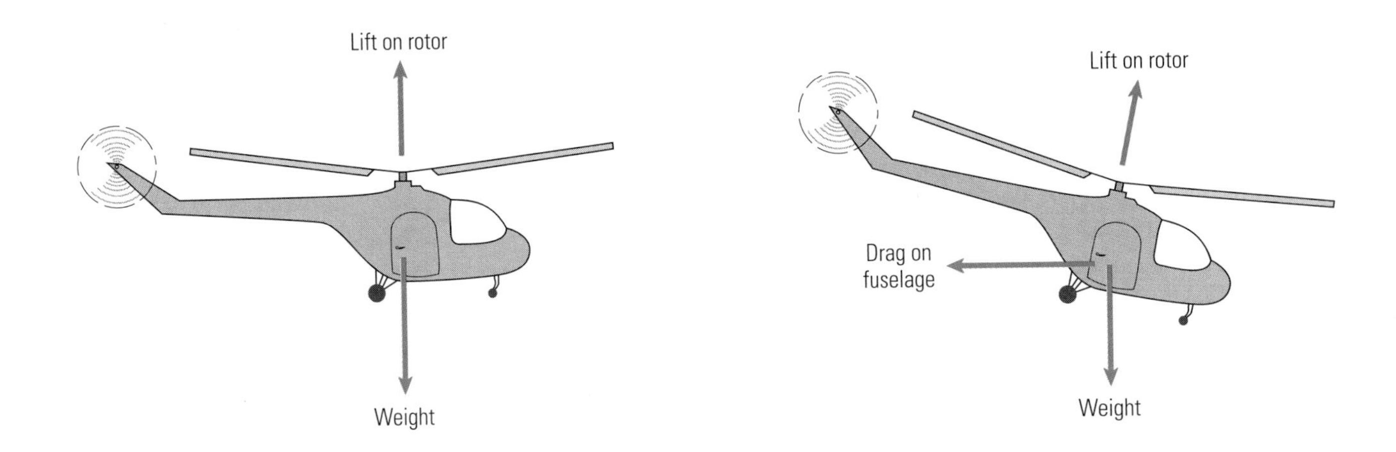


Hovering *(left)* and forward flight *(right)* of a helicopter.

An Antillean crested hummingbird (*Orthorhynchus cristatus*) hovering while it sucks nectar from a flower. The wings have just been turned upside down for the start of their backward stroke.

is kept horizontal; the air is driven vertically downward and the lift acts vertically upward. To travel horizontally, the rotor is tilted so that the air is driven at an angle, downward and backward. Lift then acts upward and forward, providing thrust as well as support.

If flying animals are indeed like helicopters, and their wings play the role of rotors, we might expect them to hold the wings horizontally while hovering and to tilt the wings at an angle while moving forward—and in fact many of them do. Hummingbirds hover like tiny helicopters as they suck nectar from flowers. The body is tilted up at a steep angle that positions the wings to beat horizontally. Instead of rotating like helicopter blades, however, the wings beat

A honey bee *(Apis mellifera)* hovering. The wings are making their forward stroke: they will turn upside down for the backward one.

A wing of a damselfly *(Calopterix splendens)*, and a small part of its pleated surface greatly enlarged in a scanning electron micrograph.

backward and forward, changing their angle at the end of each stroke to provide the angle of attack needed to give upward lift. In each forward stroke the wing faces what seems to be the right way up: the surface that is uppermost in the flight of other birds faces upward, and the anterior edge (where the bones are) is in front. However, high-speed films show that the wing turns upside down for the back stroke, so that the anterior edge faces backward. In both the forward and the backward stroke, the relatively thick, bony anterior edge of the wing is always in the lead with the feathers trailing behind. High-speed film is needed to see this because the wings beat at very high frequencies, between 15 cycles per second in large, 20-gram hummingbirds and 60 cycles per second in small, 2-gram ones.

Moths, bees, and many other insects hover in the same way as hummingbirds. In insects, too, the stiffer edge of the wing always leads, and insects also turn their wings upside down for the backward stroke. Insect wings are not stiffened by bone, however, but by veins and by pleating. Like the bones in hummingbirds, the largest veins and the deepest pleats are found near the anterior edge. Hummingbirds are alone among birds in using this technique to hover, but we need not be surprised that in this respect hummingbirds behave like insects. They are, after all, the smallest of all birds, similar in size to the largest beetles and moths.

Most birds that can hover (and only small birds can) move their wings in a very different way, more like the tit in the illustration. Notice how the primary wing feathers (the large feathers in the outer part of the wing) are bent upward during the forward stroke, showing that large lift forces are then acting on them. In the backward stroke, however, the wings are partly folded and the feathers are not bent, which shows that any lift forces are small. When tits and most other birds hover, they seem to produce lift only in the forward stroke of the wings. The wings are tilted in the forward stroke at an angle of attack that produces upward lift, but are angled to give little or no lift in the backward stroke, and they do not turn upside down.

This explanation of hovering flight may seem adequate, but problems emerge when zoologists try to calculate the forces on the wings. If you know the area of a wing and its speed of movement, you can estimate the greatest lift it can produce, by using the conventional aerodynamics that engineers apply to airplane wings and helicopter

A great tit hovering.

rotors. This is the lift that the wing would give if tilted almost to the angle of attack at which it would stall. The maximum possible lift, estimated in this way, increases and decreases in the course of each wing beat as the wing speeds up and slows down, but it can be averaged over a wing beat cycle. The calculations are quite complicated but lead to a clear conclusion: many hovering birds and insects cannot (in theory!) hover. The calculations seem to tell us that the wing movements of a hovering flycatcher (a small bird) can produce no more than one third of the force needed to support the bird's weight, and similarly that the wings of a hovering hoverfly can support only one third of the weight of that small insect.

Since flycatchers and hoverflies do in fact hover, the theory must be wrong. The explanation is that the theory was developed for airplane wings and helicopter rotors that move steadily through the air. It does not work well for unsteady movements like those of the flapping wings of birds and insects, which keep stopping and starting.

To understand how stopping and starting can affect the lift on wings, we need to know more about how air flows around them. We saw in Chapter 4 how air flowing over a stationary wing is deflected downward if the wing has a positive angle of attack. The air traveling over the top of the wing speeds up and the air going under slows down: similarly, if two people run side by side around a bend, the one on the outside of the bend has to go faster. Now, instead of thinking of air moving over a stationary wing, we will think of a wing moving through stationary air (except where the air is set in motion by the moving wing). The flow of air *relative to the wing* is the same as before, but as seen from the ground the bulk of the air is now stationary, the air immediately over the wing is moving backward, and the air under the wing is moving forward: there is a circulation of air around the wing.

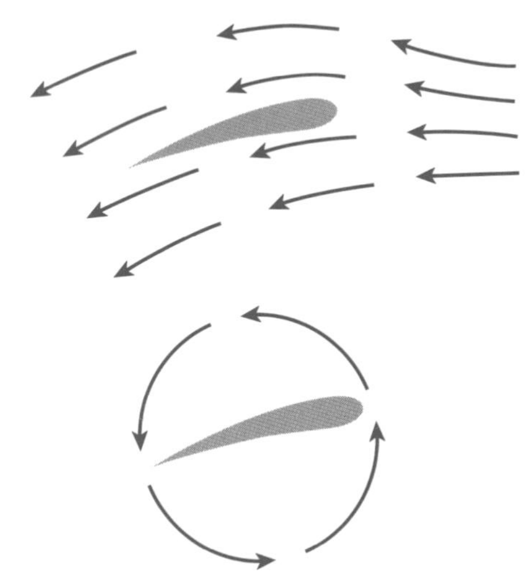

A wing moving through initially still air. The diagram above shows air movements relative to the wing, and the one below shows movements relative to the ground.

This circulation takes a little time to develop after the wing starts moving, and it does not cease immediately when the wing stops. For that reason, there is less lift on an accelerating wing than you might expect from the wing's speed, and more on a decelerating one. Birds are able to exploit unsteady effects such as these to gain more lift than conventional aerodynamics seems to allow.

One way to achieve additional lift involves clapping the wings together at the top of the upstroke. Pigeons often do this when they take off, producing quite a loud clapping noise, but the "clap and

A "clap and fling" used by a hovering insect. *Encarsia,* described in the text, has two pairs of wings, but only one pair is shown in this diagram. Thick arrows represent air movements.

fling" mechanism was first demonstrated in research on a tiny insect. Torkel Weis-Fogh of Cambridge University had taken a high-speed film of *Encarsia,* a miniature wasp with wings only 0.6 millimeter long. It hovers rather like a bee or a hummingbird, holding its body vertical and beating its wings in a horizontal plane. The wings are clapped together back to back, then flung apart, separating first along their upper (anterior) edges. Air rushes into the expanding space between the separating wings, with the result that by the time separation is complete, there is a circulation of air around them. This circulation is much stronger than could be achieved if the wing were moving steadily at the same speed instead of stopping and starting, and the lift is correspondingly greater. This is only one of several tricks that flying animals use to exploit unsteady effects and improve their lift. Dragonflies and hoverflies seem to depend for their hovering flight on different unsteady effects, moving their wings quite differently from how other hovering insects move theirs.

## Vortices and Airflow

When helicopters hover, stationary in the air, they produce the upward force that is needed to support their weight by driving air downward. To fly forward they must drive air backward as well, to supply forward thrust. Similarly, animals that are flying forward rather than merely hovering must drive air both downward (for weight support) and backward (for thrust). The air movements in the wake of a flying animal can tell us about the forces it exerts, much as force plates tell us about the forces on the feet of runners. We are going to examine the air movements set up by flying animals because they will help us to understand their flight, but first we need some general information about air movements, starting (rather surprisingly, perhaps) with smoke rings.

Blowing smoke rings was a favorite trick of smokers, in times when smoking was socially more acceptable than now. They would fill their mouths with smoke and then blow it out in a single, gentle puff. A ring of smoke formed and could be watched drifting slowly away. The creation of the ring depends on a property of moving air: as a puff of air moves through still air, it sets a surrounding belt of air rotating as a "vortex ring." (Similarly, a crate being pushed along on rollers makes the rollers rotate.) Most of a mouthful of smoke is soon dispersed, but some is trapped in the core of the vortex ring by the air rotating around it, and the ring becomes visible.

Vortices form whenever a body of air moves through still air (or through air with a velocity different from its own). Huge vortex rings form around rising thermals, for example. A whole complex of vortices forms around the air that is driven downward behind the wings of an airplane, outlining the shape of a rectangle. The two long sides of the rectangle consist of vortices that have formed around the wing

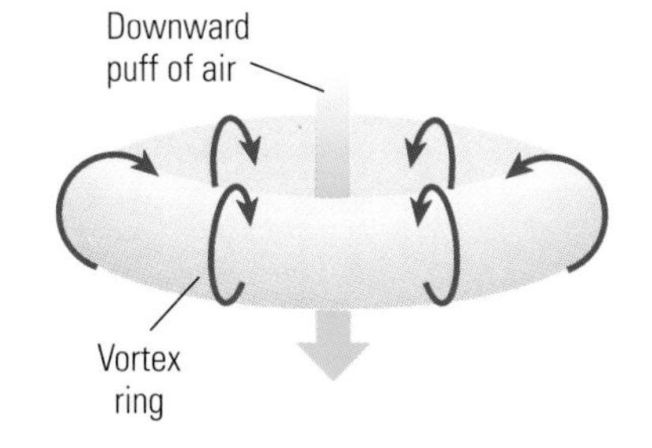


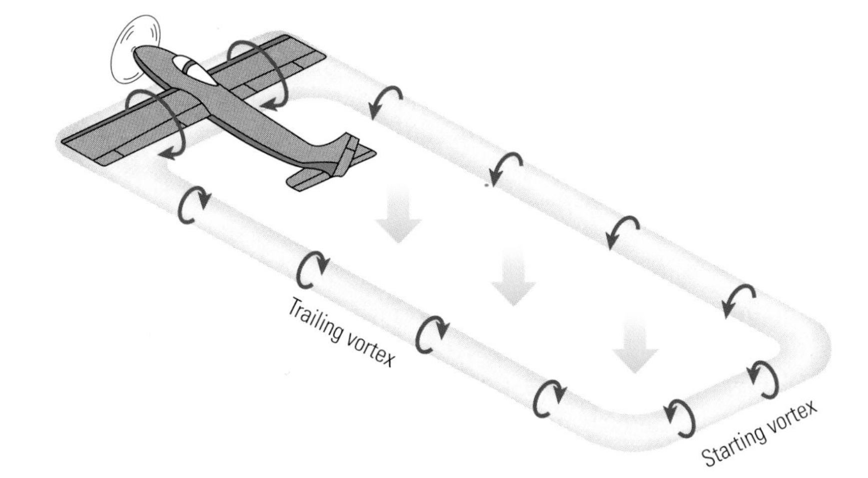


A vortex ring *(left),* and the vortices behind an airplane *(right).* Arrows represent air movements.

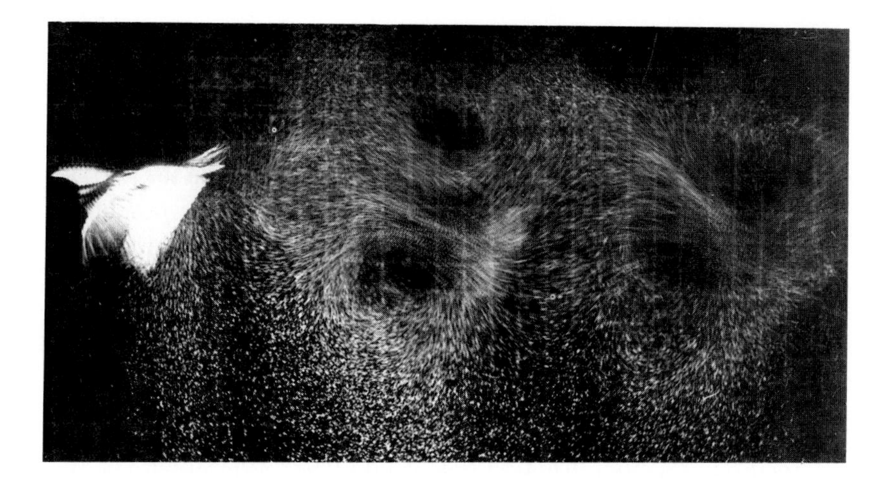

Two vortex rings appear behind a chaffinch (*Fringilla coelebs*) flying through a cloud of dust. A series of flashes were fired while the camera shutter was open, so particles in the moving air of the bird's wake appear as streaks.

tips and now trail behind; they sometimes become visible as vapor trails. A "starting vortex" running across the wake is left behind where the airplane took off. Finally, the rectangle of vortices around the downward-moving air is completed by the circulation around the wing itself.

When a hummingbird hovers, each wing stroke drives downward a puff of air, and that small air mass is presumably surrounded by a vortex ring. Successive wing beats should build up a stack of vortex rings, one on top of another, and the wake should be little different from the wake below a hovering helicopter, which produces a continuous stream of air rather than a series of puffs. No one has actually studied the airflow under a hovering hummingbird, but it seems pretty clear what it must be like.

It is less obvious what the pattern of airflow should be like in forward, flapping flight, but Nikolai Kokshaysky of Moscow obtained photographs of vortex rings by flying finches through clouds of brightly lit sawdust. His photographs seemed to show that vortex rings are produced only in the downstroke, but they could not give very accurate information about airflow because the sawdust particles did not simply move with the air, but sank through it. Many photographs of birds flying slowly or hovering show the main wing feathers bent only in the downstroke. Like Kokshaysky's photographs of vortex rings, these suggest that lift is obtained only in the downstroke.

Colin Pennycuick devised a much more sophisticated version of Kokshaysky's experiment and assembled a team in Bristol to perform it. His colleagues were Jeremy Rayner, who had devised a vortex theory of bird flight, and Geoff Spedding, then a graduate student, who did most of the practical work. They set out to make the vortices visible by having a bird fly through a cloud of tiny soap bubbles.

To obtain soap bubbles that would not sink or rise, Pennycuick's team had to create bubbles that were exactly the same density as air. Soap solution is denser than air, so air-filled bubbles sink. Helium, however, is less dense than air, and helium-filled bubbles rise. The experimenters filled their bubbles with a mixture of air and helium, in just the right proportions for them neither to sink nor to rise.

Spedding set up a machine that would blow soap bubbles of this composition, and he filled a laboratory with clouds of bubbles. The room was darkened, and a bird was let loose. As the bird flew through the bubbles, two cameras took flash photographs of the wake behind it. The cameras were angled to give a stereoscopic effect so that the positions of bubbles could be recorded in three dimensions. During a

A Tawny owl *(Strix aluco)* flying slowly through a cloud of helium-filled soap bubbles, producing a series of vortex rings.

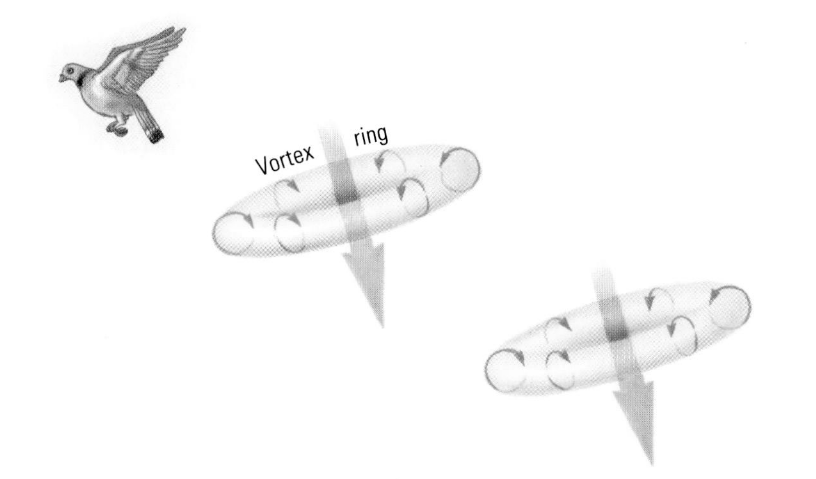


The pattern of vortices behind a pigeon flying at about 3 meters per second.

single opening of the shutter, four flashes fired in succession; thus each photograph showed four images of each bubble, indicating its path and speed of movement. Conveniently, the flashes died away rather slowly, so that each image had a tail to it; otherwise it would not have been clear which was the first, and which the last, of the four images of each bubble.

The resulting photographs show huge numbers of bubbles and are difficult to interpret, but the team managed to make sense of them. Photographs of a pigeon and a jackdaw (a European crow) verified the predictions of Rayner's mathematical theory. Each downstroke of the wings produced a vortex ring (showing that lift was being produced), but the upstroke did not.

The wings of the pigeon beat forward and down at an angle of 45 degrees to the horizontal. The vortex rings form behind the moving wing tips, but are tilted at only about 10 degrees to the horizontal. The angle of the vortex rings is so much shallower because the rings are already moving downward while they are being formed. The back of the ring, formed behind the wing tips while the wings are at the top of their stroke, has moved down some distance by the time the front of the ring is completed.

The downward flow of air follows the tilt of the rings and is thus inclined at 10 degrees to the vertical. This air is being driven downward and a little backward, so the aerodynamic forces on the bird act upward and a little forward, supporting the bird's weight and providing thrust.

The vertical component of this force can be calculated from the size of the vortex rings and the velocity of the air moving through them. Spedding measured and calculated as necessary, but reached a disturbing conclusion. He found that this supporting force, averaged over a wing beat cycle, was less than the bird's weight. An error in the calculations was suspected, but none could be found. The most likely explanation seems to be that at the stage of their flight when they were photographed, the birds were not fully supporting their weight, but were to some extent falling.

We can think of the force on the wings as lift and drag, as we did when we analyzed gliding, the lift acting perpendicular to the direction of movement of the wings and the drag acting backward along their path. Lift acts in the downstroke only, in an upward and forward direction, and its forward component is sufficient to overcome the drag on the body.

The observations of pigeons and crows confirmed expectations in showing a series of vortex rings, one for each downstroke, but experiments with a kestrel brought a surprise. Its wake consisted not of vortex rings but of a pair of continuous vortices trailing from the wing tips. These were not straight like the wing tip vortices of airplanes but wavy, rising and falling with the beat of the wings. Moreover, the vortices were farther apart for the downstroke (when the bird's wings were fully spread) than for the upstroke (when the elbow and wrist joints in the wings were flexed a little, reducing the wing span).

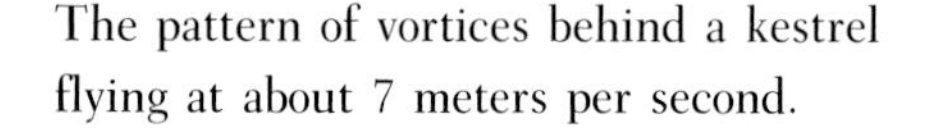

The pattern of vortices behind a kestrel flying at about 7 meters per second.

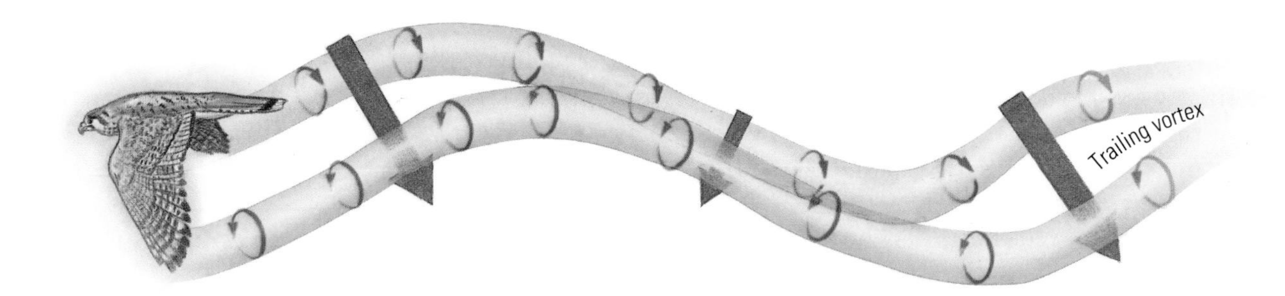

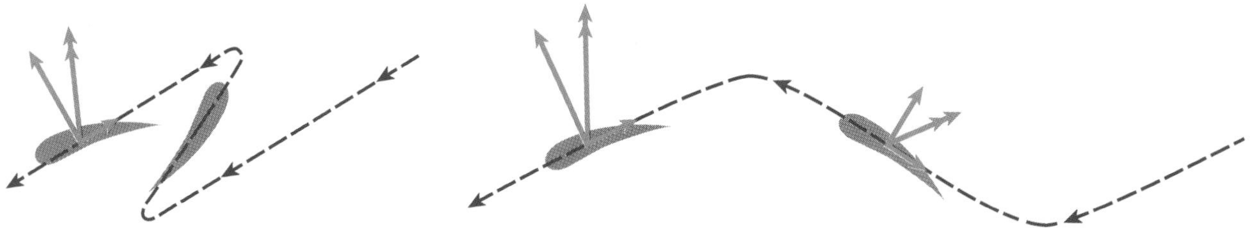

Aerodynamic forces on the wings of a pigeon flying slowly *(left)* and a kestrel flying fast *(right)*.

The downward airflow between the wing tip vortices is everywhere at right angles to them, downward and a little backward during the downstroke but downward and a little forward during the upstroke. The wings are pushing down on the air, getting the upward force needed for support, in both strokes. However, they are providing thrust only during the downstroke: during the upstroke the horizontal component of force acts in the opposite direction. If the bird is to maintain its speed, the thrust produced in the downstroke must not be cancelled out by this retarding force in the upstroke. This unfortunate effect is avoided by making the aerodynamic forces smaller during the upstroke than during the downstroke. The bird pushes strongly on the air during the thrust-producing downstroke and more gently in the retarding upstroke, so the net effect over a wing beat cycle is support plus adequate thrust.

The forces during the upstroke could be reduced by one of two means: either by accelerating the same amount of air as in the downstroke, but to a lower speed, or by driving less air at the same speed. The latter is what happens. Air flows downward between the wing tip vortices at the same speed in both strokes, but less air is moved in the upstroke because the wings are less widely spread, bringing the vortices closer together.

Again, we can think of the forces on the wings in terms of lift and drag. During the downstroke, the path of the wing is tilted downward, so lift acts upward and forward, but during the upstroke its path is tilted upward and the lift acts upward and backward. The forces on the wing must be reduced in the upstroke if the resultant of lift and drag, averaged over a complete wing beat cycle, is to have a

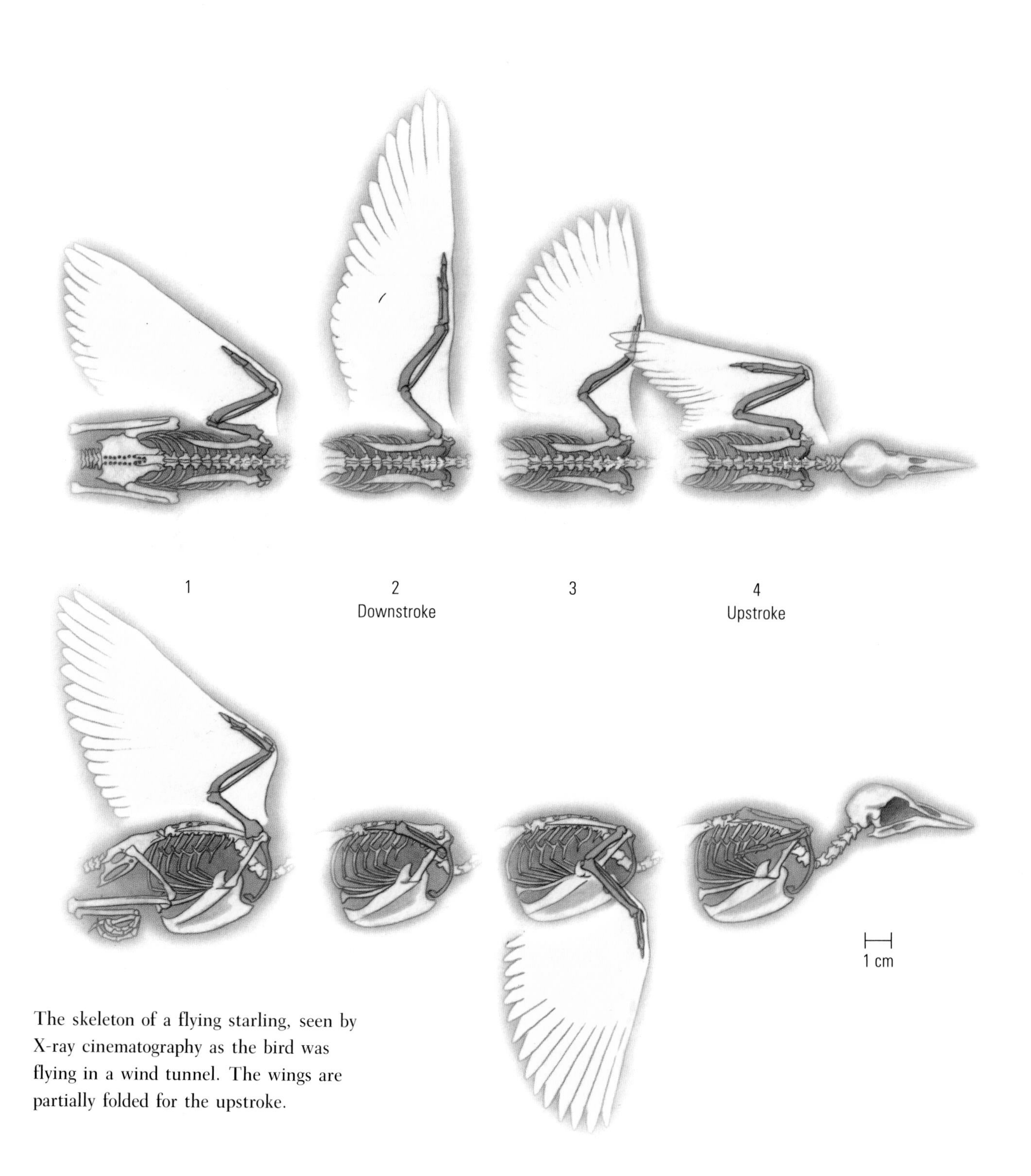


The skeleton of a flying starling, seen by X-ray cinematography as the bird was flying in a wind tunnel. The wings are partially folded for the upstroke.

forward thrust component as well as an upward weight-supporting component.

The experimental results suggested a connection between the speed of flight and the pattern of vortices. The pigeon and jackdaw flew slowly in the experiments, at about 2.5 meters per second (6 miles per hour), and produced vortex rings in the downstroke only. The kestrel flew much faster, at about 7 meters per second (16 miles per hour), and produced continuous wavy wing tip vortices. It seemed possible that birds in general might use one style of flight to go slower and the other to go faster: the two styles might be gaits used at different speeds, like walking and running. The question might have been settled if the pigeon and jackdaw had sometimes flown fast or if the kestrel had sometimes flown slowly, but (infuriatingly) the birds would not change speeds in the experimental setting. In the wild, both pigeons and jackdaws usually fly quite fast.

A noctule bat was more obliging. It usually flew slowly through the bubbles at 1 to 3 meters per second, but occasionally it went much faster, increasing its speed to 7 to 9 meters per second. When flying slowly it used the vortex ring gait like the pigeon and jackdaw, but when flying fast it used the continuous vortex gait, like the kestrel. Unfortunately, Jeremy Rayner and the students who made these observations obtained no records of intermediate speeds, so we do not know whether one gait changes gradually to the other as speed increases or the change occurs abruptly, like the change from walking to running.

At low speeds, walking uses less energy than running, whereas at high speeds the reverse is true. It seems reasonable to guess that the vortex ring gait is more economical than the continuous vortex gait at low speeds and that the continuous vortex gait is more economical at higher speeds. A simple argument suggests why this might be true, although mathematical analysis is still needed to verify the argument.

In the vortex ring gait, air is driven downward only during the downstroke, so the bird pushes on less air than if lift were being produced continuously, and consequently that air has to be accelerated to a higher speed. We have already seen that it is more economical to obtain lift by accelerating a lot of air to a low speed than by accelerating less air to a higher speed. Therefore, we would expect the continuous vortex gait to be the more economical one, but it has a characteristic that at lower speeds causes the advantage to tip toward the vortex ring gait. The characteristic is that the downstroke and

The Andean flamingoes *(Phoenicopterus ruber, left)* are flying fast, beating their wings up and down, but the Canada geese *(Branta canadensis, right)* are flying slowly, using a much more horizontal wing beat.

the upstroke drive the air in different directions (downward and back, or downward and forward). As a consequence, more kinetic energy has to be given to the air than if it were all driven in the direction of the required force. At low speeds, the animal travels less far in each wing beat cycle, so the paths of the wings in the downstrokes and upstrokes slope more steeply down and up than they do at higher speeds. If the continuous vortex gait were used at low speeds, the air would be driven in very different directions in the downstrokes and upstrokes, and so the extra kinetic energy needed because the air was not all driven in the required direction would be relatively larger. It seems likely that the vortex ring gait needs less power from the muscles at low speeds and the continuous vortex gait needs less power at high speeds.

## Bounding Flight

Many birds rise and fall as they fly, taking a wavy path through the air. Careful observation shows that there are two distinct wavy styles of flight (we can think of them as two more gaits). Crows and gulls, and also some bats, flap their wings for a few cycles, make a brief

glide with their wings spread, flap for another few cycles, and so on: this is called undulating flight. Sparrows and many other small perching birds use a different gait called bounding flight, as do woodpeckers, kingfishers, and small parrots. The wings beat for a few cycles, then are folded against the sides of the body for a while, then beat again. They beat while the bird is in each trough of its wavy flight path, and they are folded at the crests. In the case of sparrows, for example, bursts of about six wing beat cycles, lasting 0.3 second, alternate with 0.3-second periods of wing folding.

A few years ago, we had a theory of bounding flight that seemed convincing. While the wings are folded, the bird avoids the profile drag that would act on the wings if they were spread. However, if they are folded for part of the time, they must drive air down faster during the time that they are beating in order to increase the work done against induced drag. At low speeds, induced drag is large and profile drag is small, so the profile power that can be saved by folding the wings is less than the increase in induced power, but at high speeds the reverse is true. Thus birds should beat their wings continuously when flying slowly, but use bounding flight when flying fast. Mathematical analysis suggested that the change should be made at a speed just a little above the maximum range speed.

We had to discard that theory when Jeremy Rayner pointed out that many birds use bounding flight even when flying slowly. Indeed, tits, finches, and sunbirds have been observed bounding even while hovering. (Sunbirds are small birds found in the Old World tropics that behave rather like hummingbirds.)

An alternative theory of bounding flight considers the metabolic energy used by the muscles rather than the work that they do. We saw in Chapter 1 that the metabolic energy used while doing a given amount of work is least if the muscles shorten at about one third of their maximum rates. This is also the rate of shortening at which they can give the highest power outputs: muscles are most efficient and deliver most power at about the same shortening rate. Evolution matches the properties of muscles to the jobs they have to do, and it seems likely that bird muscles have evolved so that their most efficient rate of shortening, and the rate at which they can produce most power, is the rate at which they work in fast flight, when maximum power output is required. We do not yet know whether this is true of birds, but an experiment described in Chapter 7 seems to show that fish muscles have evolved according to this principle.

The alternative theory proposes that bounding flight allows birds to use their muscles at the most efficient rate of shortening. Suppose that a bird is flying at a speed that does not require its muscles to produce maximum power. In that case it has a choice between beating its wings continuously and slowly, or intermittently and rapidly. If (as seems likely) the muscles work more efficiently at the higher speed, the bird may save metabolic energy by using the intermittent gait, even if the work required is a little increased. In that case the bird should beat its wings continuously only when a very high power output is needed; for example, when climbing fast or accelerating.

If the theory is correct, a sparrow in bounding flight would use less metabolic energy than if it flew the same distance in the same time, beating its wings continuously. However, if the wings beat for only half the time in bounding flight, they would have to do the same (or slightly more) work in half the time: during the bursts of flapping, their power output must be double what would be needed in continuous flapping flight. Bounding flight is an option only for birds that have plenty of power in reserve. We will soon see that small birds have more power in reserve than large ones, which seems to explain

This photograph has caught a Scott's oriole *(Icterus parisorum)* at the stage of bounding flight at which the wings are folded.

why only small birds bound. The power needed in the flapping phase of undulating flight is less than that needed for bounding flight, because the wings are kept spread and provide lift in the intervals between bouts of flapping. Because undulating flight needs less reserve power than bounding flight, undulating flight is possible for larger birds.

Another way to reduce the energy cost of flight is to reduce the number of vortices, a strategy that may explain the elegant V formation of flying geese. Fewer vortices mean less work, because the bird does not have to impart kinetic energy to the rotating air. Birds might be able to eliminate vortices by flying in groups, if they positioned themselves correctly. For example, they might fly side by side with their wing tips touching, so that their wakes of downward-moving air merged and there were wing tip vortices only at the edges of the group. The left wing of the leftmost bird and the right wing of the rightmost bird would leave wing tip vortices, but the others would not. Birds do not fly with wing tips touching (they might have some nasty accidents if they did), but groups of geese often fly in V formation with the left wing tip of one bird, for example, close behind the

The wing tip vortices behind a formation of geese.

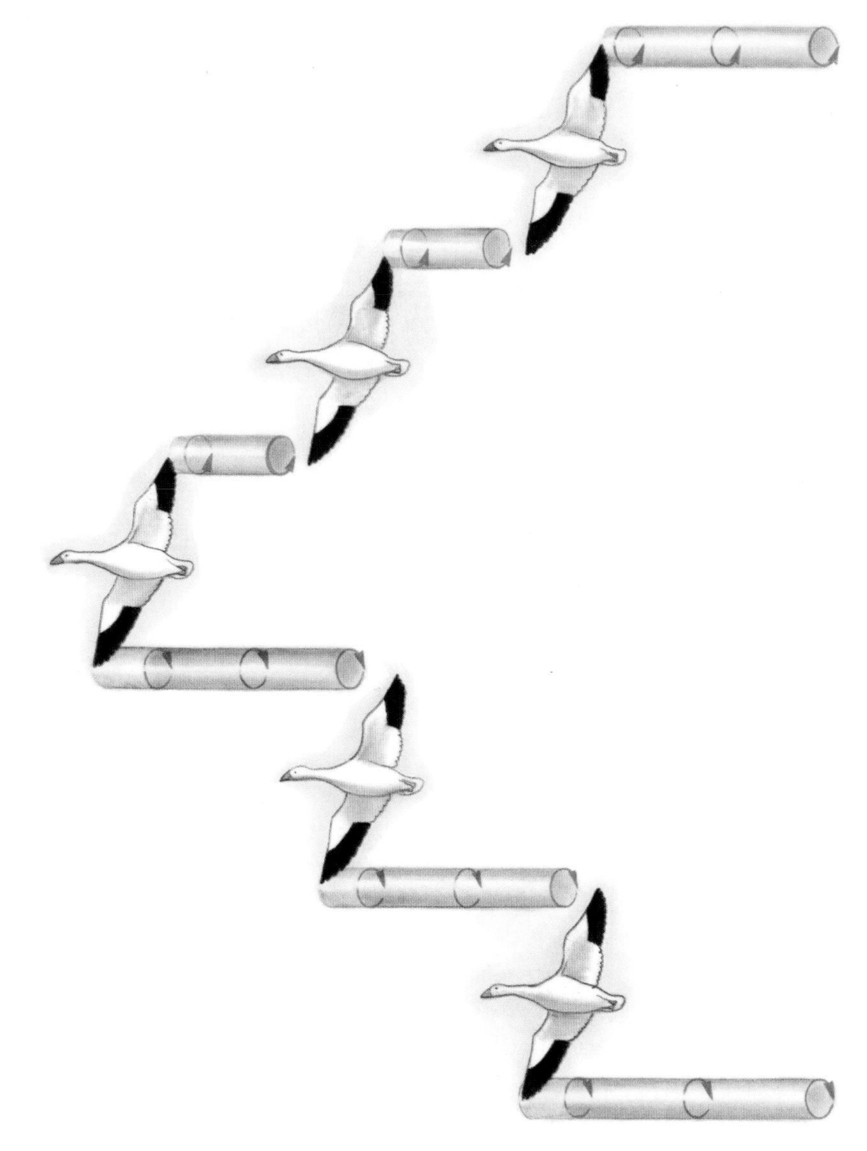

right wing of another. If the relative positions of the birds were right, the wing tip vortex of the bird in front would be cancelled by the effect of the wing tip of the bird behind. There would be vortices in the wake only behind the outer wing tips of the birds on the edge of the formation, and energy would be saved. This seems to happen.

Snow geese *(Chen caerulescens)* flying in V formation, each with a wing tip behind the wing tip of the bird in front.

## Can Bumblebees Fly?

The forward flight of insects involves patterns of force on the wings that are intriguingly different from those found in the forward flight of birds. Scientists used to be fond of saying that aerodynamics had proved that bumblebees cannot fly. The behavior of the bees themselves shows us that this was nonsense, and in any case we now understand bee flight better than we did. Let us see what recent research can tell us about how bees do fly.

The bumblebee is the best example to take in a discussion of insect forward flight because the experiments that have been done on it are far more satisfactory than those done on other species. In these other experiments, the insect's body was glued temporarily to a wire, which was then fixed so that the animal was held supported in an air current, facing into the wind. In these circumstances many species beat their wings as if they were flying. This convenient behavior has been exploited in many investigations of insect flight mechanics: insects have been filmed, or the forces on their bodies have been measured, while they "flew" fixed in wind tunnels. The results have doubtful value because the movements may not be the same as if the insects were really flying; for example, locusts that beat their wings at 23 cycles per second in free flight made only 20 cycles per second in tethered flight.

The most detailed study of flight by free-flying insects has been made at Cambridge University by Charles Ellington and his student Robert Dudley. They filmed bumblebees flying in a wind tunnel, in a space enclosed by plastic netting. If the bees were put into the empty enclosure they flew irregularly, apparently because they missed the movement of the landscape through their field of view. Dudley and Ellington persuaded them to fly steadily by placing rotating cylinders with barber's-pole stripes on either side of the enclosure. When these were rotated so that the pattern of stripes seemed to move slowly downwind, the bees flew steadily into the wind, sometimes for as long as $1\frac{1}{2}$ minutes.

Bumblebees beat their wings at 150 cycles per second, so filming with an ordinary movie camera would have been useless. (Movie film is usually taken at 18 or 24 frames per second, just fast enough to avoid a flickering effect when the film is shown.) Instead, Ellington and Dudley used a special high-speed camera running at 5000 frames per second, enough to give them about 30 pictures of each wing beat cycle. With their single camera they could not achieve a stereoscopic effect, but they were able to work out the movements of the wings in three dimensions by a method that depended on the assumption that the movements of the left and right wings were symmetrical. They filmed from an oblique angle and measured the positions of corresponding points on the left and right wings.

We have already compared the hovering flight of hummingbirds and insects to the hovering of helicopters. The aerodynamic force provided by a helicopter rotor is perpendicular to the plane in which its rotor blades move. To hover, the rotor is kept horizontal, but to fly forward it is tilted so that the aerodynamic force on it has a forward (thrust) component as well as the vertical component that balances the aircraft's weight.

Bumblebees flying in the wind tunnel behaved in essentially the same way. When hovering, they held their bodies tilted at about 50 degrees to the horizontal, and their wings beat in a horizontal plane. At a low forward speed of 2 meters per second (bees fly between closely spaced flowers at about this speed), both the body and the plane of the wings were at 25 to 30 degrees to the horizontal, and in fast flight at 4.5 meters per second, the body was held at 10 degrees and the wings beat at 40 degrees. Whatever the speed, the wings beat at about the same frequency and through the same range of angles.

Dudley and Ellington made no attempt to observe the vortices in the wake; instead, they estimated the forces acting on the wings by measuring the angles of attack. In hovering, the wings moved like hummingbird wings, backward and forward in the same horizontal plane, turning upside down for the backward stroke and presumably producing lift in both strokes. In forward flight, the insect moved ahead through the air while the wings beat up and down, so the wings followed a sawtooth path, moving through the air at different angles in the down- and upstrokes. This implies that the forces on the wings act at different angles in the two strokes. In fast flight the downstroke seems to supply most of the vertical force needed for weight support, and the upstroke most of the forward thrust. The wing moves faster relative to the air in the downstroke, because the forward velocity of the wings relative to the body is then added to the forward movement of the body through the air. Thus, we can expect aerodynamic forces to be larger in the downstroke than in the upstroke. The fast flight of the bee seems to be quite different from any gait of birds, which do not get thrust from their upstrokes.

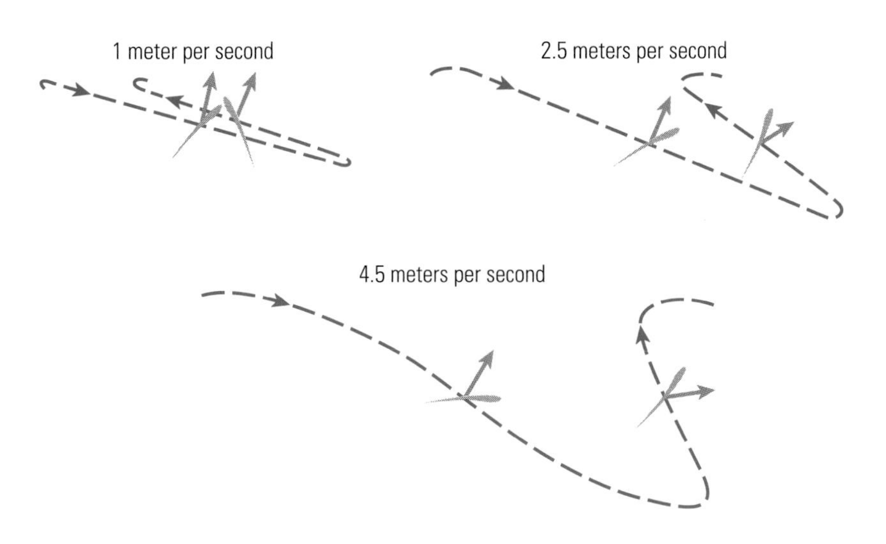


The path of the wings of a bumblebee flying at different speeds, and the aerodynamic forces on them. Each diagram shows two sections through the wing, one during the downstroke and one during the upstroke.

## Wings to Suit Flying Habits

Birds, bats, and insects each have their own distinctive kind of wings: bird wings consist of feathers attached to the bones of a modified arm and hand; bat wings are made of skin stretched between immensely elongated fingers; and insect wings are thin sheets of cuticle stiffened by thickenings known as veins and by light pleating. Although each group conforms to its basic wing plan (for example, no insect or bat has evolved feathers), variations on the basic plan have evolved within each group to suit different flying habits. We have already seen that albatrosses have long but very narrow wings and that vultures have much broader ones: albatrosses have much higher wing loadings and aspect ratios than vultures of the same body mass. Similarly, the broad wings of a butterfly are very different from the narrow wings of a dragonfly. The relationship of body structure to flying habit is brought out sharply in a study by Jeremy Rayner.

Rayner studied the sizes and shapes of bird wings, using the statistical technique known as principal components analysis. He

A horseshoe bat (*Rhinolaphus ferrum-equinum*) chasing a moth. Three flashes fired in rapid succession captured three images as the bat flew past.

This triple-flash photograph of a lacewing (*Chrysopa carnea*) taking off from a leaf shows the two pairs of wings beating out of phase with each other. Notice how the flexible wings bend and twist in the course of their stroke.

started with measurements of body mass, wing area, and wing span for hundreds of different species. Plainly, these measurements are related to each other (heavier birds generally have larger wings), but the relationships are not simple. If birds were geometrically similar to each other, wing areas would be proportional to the squares of the birds' lengths and masses to the cubes of the lengths, implying that wing areas would be proportional to the two-thirds power of body mass—four times the area for eight times the mass. Actual relationships are quite close to that estimate for some groups of similar birds of different sizes (for example, petrels and albatrosses), but we could not have assumed that they would be.

The principal components analysis automatically reorganizes the data, calculating quantities called "principal components" that vary as independently of each other as possible. The first principal component obtained in Rayner's analysis of bird wings simply shows whether each bird is large or small, basing its measure of size not only on mass, but also on wing area and span. The second component tells whether the bird has high or low wing loading, *compared to other birds of similar size.* The third tells whether its aspect ratio is high or low, compared to other birds of the same size and wing loading.

A graph shows the second and third components, the ones that are related to wing loading and aspect ratio. Different groups of birds occupy different parts of the graph, according to their different flying habits. Most seabirds have high aspect ratios and are placed in the upper half of the graph: gannets, albatrosses, gulls, terns, and others are all found there. However, among seabirds, frigate birds (which soar on thermals) are well over to the left, with low wing loadings, and murres, or guillemots (which use their wings to swim underwater, as well as for flight), are well over to the right, with high wing loadings. Birds that soar in thermals over land (vultures, pelicans, storks, and others) lie in or near the lower left quarter of the graph: they have low wing loadings and low aspect ratios, as we have already seen. Birds of prey (owls, hawks, kites, and eagles) have low wing loadings, so they can still fly well when their wings have to support a load of heavy prey. Game birds (grouse, pheasants, peacocks, and turkeys) have small, broad wings and do not fly well; they are found in the lower right quarter.

Though birds with similar flying habits are generally grouped near each other on the graph, groups with different habits sometimes mingle. The swifts and swallows, which catch insects in the air, are mixed in with the seabirds because they too have long, narrow wings. The hummingbirds are quite near the diving birds because both have rather small, narrow wings, though for quite different reasons.

A patch of color shows how bats fit into the analysis: some have narrow wings and some rather broad ones, but all have rather large wings for their size. Differences among bats (not shown on the graph) reflect their different habits. For example, the few bats that hunt for frogs, small mammals, or fish have lower wing loadings than most other bats, enabling them to carry heavy prey; but those that hunt for frogs or mammals among vegetation have shorter (lower aspect ratio) wings than those that fish.

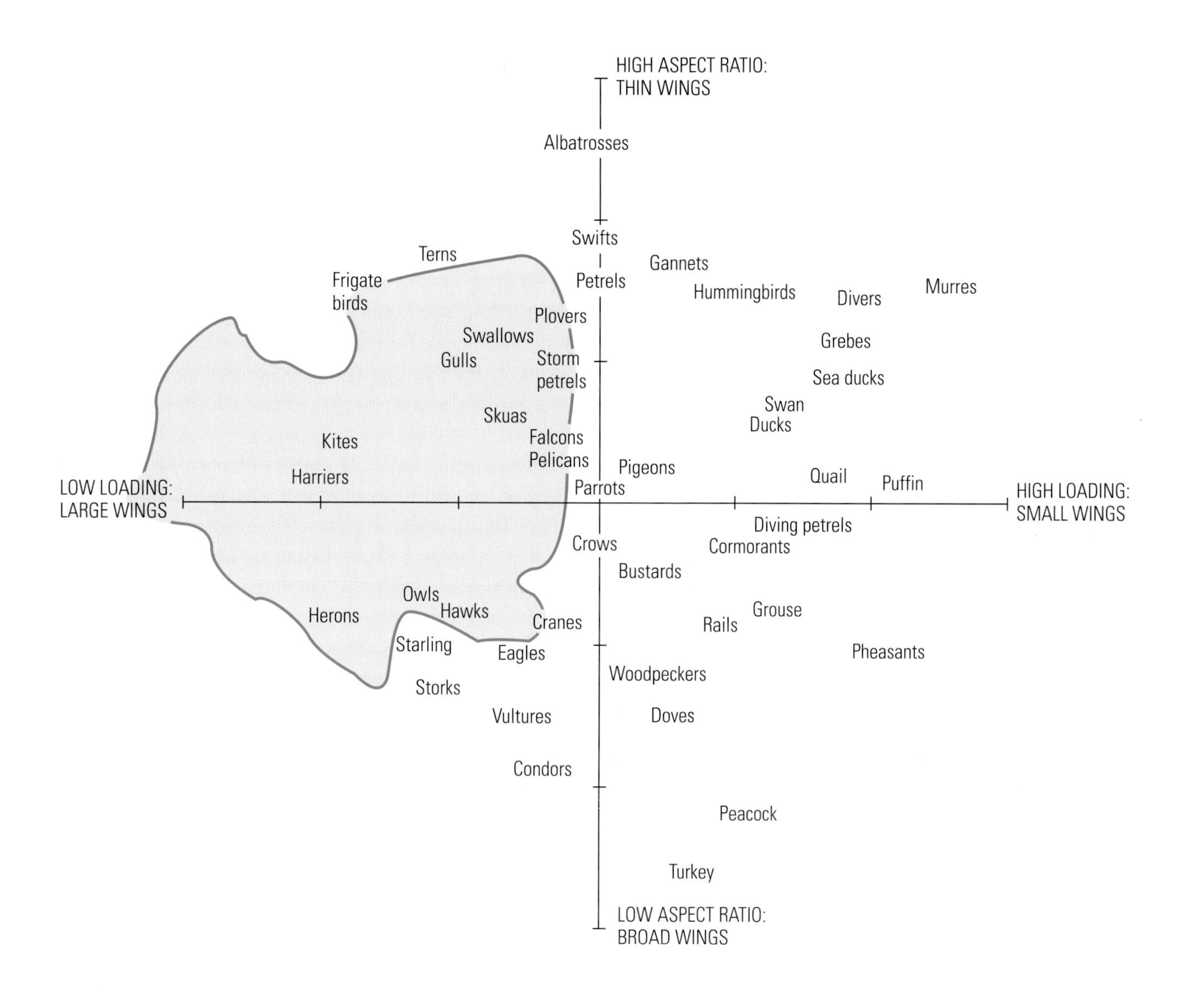


Principal components analysis of wing dimensions separates birds with different flying habits. The colored patch shows the area occupied by bats.

Birds with different habits may have very different wing muscles. Chickens, which fly only in short bursts, have white anaerobic muscles to beat their wings, whereas hummingbirds, which fly for long periods, have much darker aerobic muscles. In most birds the supracoracoideus muscles, which raise the wing, have only about one tenth the mass of the pectoralis muscle, which powers the downstroke: there is no need for large wing-raising muscles because lift helps to stop the wings at the end of the downstroke and (in fast flight) to raise them. In hummingbirds, however, the supracoracoideus has to power the backward wing stroke in hovering and has half the mass of the pectoralis.

Animals of different sizes beat their wings at very different frequencies: 2 to 2.5 wing beat cycles per second for 10-kilogram condors, about 20 cycles per second for 10-gram hummingbirds, 260 cycles per second for very large, 10-milligram mosquitoes, and even more for small midges. You can count the slow wing beats of a condor (if you are lucky enough to see one); you cannot see the individual wing beats of a hummingbird, but their frequency is high enough that you can hear a low hum; and you can hear a high-pitched buzz from a mosquito.

It may seem obvious that condors cannot flap their wings fast enough to buzz, but let us think about the reason why. Imagine two geometrically similar flying animals, one 10 times longer than the other and 1000 ($10 \times 10 \times 10$) times heavier. The larger one has 1000 times more wing muscle than the smaller one, able to do 1000 times more work to accelerate the wings at the beginning of each stroke. Work is needed to give the wings kinetic energy, $\frac{1}{2} \times \text{mass} \times (\text{speed})^2$. But even though 1000 times the work is available, the wings have 1000 times the mass. The two factors cancel each other, and the larger animal can accelerate its wings to only the same speed as the smaller one.

To work out the effect of size on wing beat frequency, we next consider the span of the wings. The large animal has wings 10 times longer than the small one, so their tips have to travel 10 times farther than the tips of the small one, if they beat through the same angle. Thus, if the wing speed is the same as for the small animal, the wings will take 10 times longer to go through a cycle, and the wing beat frequency is divided by 10. This argument (which is admittedly oversimple) says that if a 10-milligram mosquito beats its wings at 260 cycles per second, a 10-gram hummingbird (1000 times heavier) should beat at 26 cycles per second and a 10-kilogram condor (a further 1000 times heavier) at 2.6 cycles per second. These frequencies are not far from the truth. A similar argument explains why elephants take fewer strides per second than mice.

## Wings That Oscillate

The 260-cycle-per-second wing beat frequency of a mosquito is high, but other small insects beat their wings even faster, at up to 1000 cycles per second in a midge. This amazingly high rate would not be possible if the insects did not have special wing muscles. Those insects whose wing muscles have properties like those of human muscles, though capable of shortening faster, can beat their wings at frequencies up to only 100 cycles per second. Flies, bees and wasps, beetles, and the insects that are properly known as bugs have a very special kind of muscle called fibrillar flight muscle, and some of them have very much higher wing beat frequencies.

It became clear that there was something odd about fibrillar flight muscle when electrodes were put into the wing muscles of tethered flies to record the electrical signals (called action potentials) that travel through muscles when they are stimulated by their nerves and make them contract. Ordinary muscles need an action potential to start every contraction, but these muscles were passing only three action potentials every second, even though they were making 120 contractions each second to beat the wings—there was only one action potential for every 40 contractions. It seemed that the muscles were somehow going into a state of oscillation that needed only occasional stimuli to sustain it.

The nature of the oscillations was made clearer by an ingenious experiment performed by Ken Machin and John Pringle at Cambridge University. They wanted to avoid the special problems of experimenting with tiny muscles, whose high frequencies of oscillation would be difficult to observe, and for that reason they used the largest insects they could get, rhinoceros beetles and giant water bugs up to 11 centimeters long. Their experiment was equivalent to connecting the wing muscle to a tuning fork and then stimulating the muscle to contract. A tuning fork has a resonant frequency of vibration—when struck it will vibrate at that frequency, emitting its particular musical note. The vibrations of a struck tuning fork gradually die away, but a muscle capable of very rapid contractions could start a fork vibrating and *keep it vibrating indefinitely* by making a contraction in each cycle of the fork's resonant vibrations. In Machin and Pringle's experiment, however, the muscle did not sustain the vibrations of a simple fork, but of a complex electronic device that had a resonant frequency that could be altered as the experimenters chose. They found that a muscle from a beetle, which would in life have made 40 wing beat cycles per second, could be made to work at frequencies at least from 30 to 70 cycles per second by adjusting the resonant frequency of the apparatus. There was no need to change the frequency of the electrical stimuli: the muscle automatically matched its frequency of contraction to the resonant frequency of the apparatus. The muscle would be capable of driving tuning forks of a wide range of frequencies, each at its own specific frequency.

To appreciate the significance of Machin and Pringle's experiment we need to know more about resonant systems. Any system that has mass and elastic compliance (meaning that it can be distorted and will spring back elastically) has a resonant frequency and can be made to vibrate at that frequency much more easily than at any other. To demonstrate this, take a rolled magazine and attach to it a chain of long, thin rubber bands. The demonstration will work well if the weight of the magazine is enough to stretch the rubber bands by about 25 centimeters (10 inches). Take the free end of the chain of rubber bands in your hand and hold the magazine suspended. Now move your hand repeatedly up and down through a distance of about 5 centimeters (2 inches). If you do this very slowly, the magazine will mimic the movements of your hand, rising as it rises and falling as it falls, through about 5 centimeters. If you do it very fast, the magazine will remain almost stationary as your hand moves up and down. However, there is an intermediate frequency (about 1 cycle per second if you perform the experiment exactly as suggested) at which small up and down movements of the hand will cause larger up and down movements of the magazine than at any other frequency. This is the resonant frequency, and its value depends on the mass and the elastic compliance. If you use two copies of the magazine, doubling the mass, the frequency is reduced by about 30 percent. If you use twice as long a chain of rubber bands, doubling the compliance, the frequency is also reduced by 30 percent.

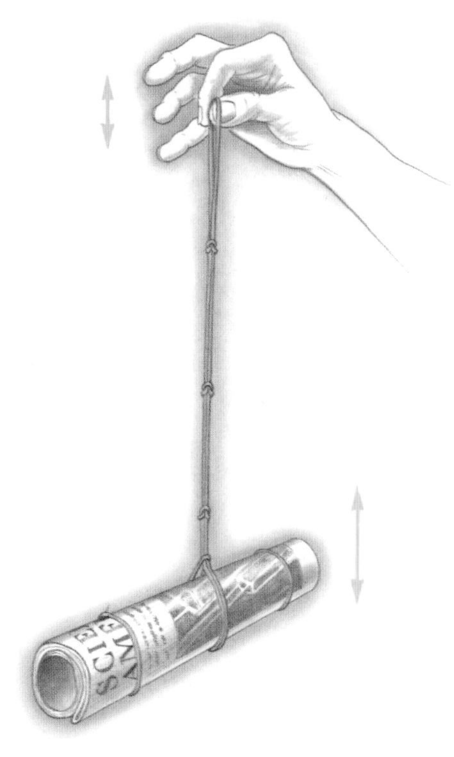

Resonant systems, such as this one of a magazine hanging by a chain of rubber bands, can be made to vibrate at their natural frequency much more easily than at any other.

Machin and Pringle's experiment showed that the body of an insect with oscillating wings contains a resonating system similar in principle to the magazine and rubber bands. Insect bodies have three parts: the head, the thorax to which the wings and legs are attached, and the abdomen. The wings and thorax together have a resonant frequency at which they can be made to vibrate most easily. The most important mass in this system is the mass of the wings: they are only a tiny fraction of the total mass, but it is their mass that matters because they are the moving parts. The elastic compliance is partly in the cuticle of the thorax wall and partly in the muscles themselves. The importance of resonance in setting the frequency of oscillation is very easily demonstrated by a simple experiment. If the wings of a tethered fly are cut short, reducing their mass, they beat at a higher frequency. If, on the other hand, they are loaded with small weights, their frequency of beating is reduced. In the intact fly, just as in Machin and Pringle's experiment, the muscles adjust their frequency of contraction to match the resonant frequency of the system that they are driving.

The thorax could be constructed as a nonresonant system. In that case the muscles would have to do work to accelerate the wings at the beginning of each stroke and negative work (acting as brakes) to stop them at the end. The quantities of work would be large: Charles Ellington's calculations for various insects show that the positive work needed to accelerate the wings is typically about twice the work that has to be done against aerodynamic drag. If the system is resonant, this extra work is not required because as the wings are stopped at the end of a stroke, their kinetic energy is converted to elastic strain energy in the muscles and thorax wall. Later, this energy is returned in an elastic recoil for the next stroke. Thus energy is saved in much the same way as it is saved by tendon elasticity in running. The resonance of the thorax has two important functions: it saves energy, and it enables fibrillar flight muscles to operate at extremely high frequencies in small insects.

No such resonant system has been found in birds and bats. Their wings beat at frequencies far below the range at which fibrillar flight muscles are needed, but there does seem to be scope for energy savings. Torkel Weis-Fogh estimated that if hummingbirds had resonant wings, the work that their muscles have to do when they hover would be reduced by 43 percent.

## The Hard Work of Flying

Imagine yourself with wings, trying to fly. You probably feel that it would be very hard work, and you are right. We will make a rough calculation of the energy cost of flight and then go on to see how it has actually been measured.

Gliding is flight powered by gravity. We saw in Chapter 4 that gliding birds and bats lose height at rates between about 1.0 and 2.5 meters per second. As they descend, they lose an amount of potential energy equal to (mass × gravity × height change). A 1-kilogram mass descending 1 meter loses (1 × 10 × 1) = 10 joules potential energy, so gliding birds lose potential energy at rates between about 10 and 25 joules per second (that is, between 10 and 25 watts) for every kilogram of their mass. The lost potential energy is the power that propels the birds through the air, doing work against drag. In level flight the power has to be supplied by muscles that are unlikely to work with better than 25 percent efficiency—4 joules of metabolic energy must be used to perform 1 joule of work. Thus we can expect the metabolic energy cost of flight to fall between 40 and 100 watts per kilogram.

The actual power has been measured by experiments with animals flying in wind tunnels. Vance Tucker of Duke University was the first to make such a measurement. He trained budgerigars (small Australian parrots) to fly in a wind tunnel wearing masks connected to oxygen analysis equipment: this was the aerial equivalent of Richard Taylor's measurements of the oxygen consumption of mammals running on moving belts. When they were resting, the budgerigars used 20 watts of metabolic energy per kilogram of body mass, but when they flew at 10 meters per second they used oxygen six times faster, at a rate corresponding to 120 watts per kilogram. The difference of 100 watts per kilogram (shown in the graph) is the energy cost of flight and is at the top of the range suggested by our initial calculation. The birds used more power when they flew more slowly and also when they flew faster. This had been expected from studies of aircraft flight, because the power needed to propel an aircraft is high at low speeds (when induced power is high) and high at high speeds (when profile power is high), and is lowest at an intermediate speed, the minimum power speed.

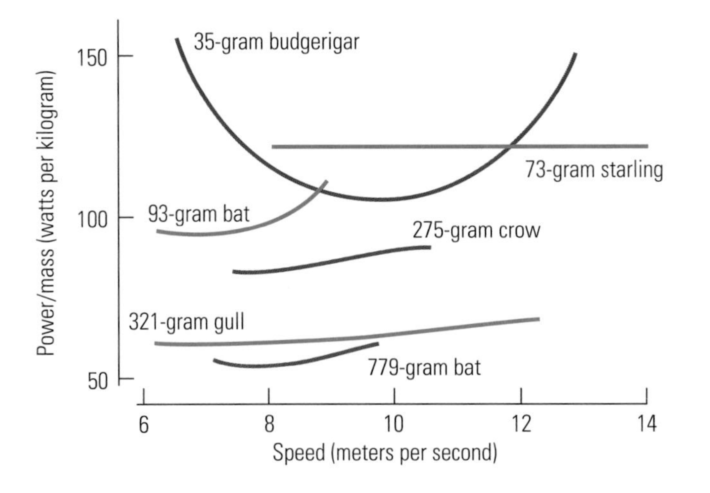


The metabolic energy cost of flight at different speeds, for some birds and bats. The resting metabolic rate has been subtracted from the rate during flight to obtain the power (rate of energy use) required for flying.

Since Tucker performed his pioneering experiment, he and others have made similar measurements on various birds and bats. They have found metabolic energy costs ranging from about 50 watts per kilogram (for an 800-gram bat) to 120 watts per kilogram (for 70-gram starlings), in good agreement with our rough calculation. Budgerigars and pigeons used less power at a moderate speed than when flying faster or slower, as expected, but starlings, crows, and gulls gave flat graphs showing little change of power consumption with speed. So far, this anomoly has not been explained satisfactorily.

Charles Ellington and colleagues at Cambridge have recently succeeded in measuring the oxygen consumption of bumblebees. Many measurements had previously been made of the oxygen consumption of tethered insects beating their wings, but these measurements were the first for insects in free flight. Insects breathe through spiracles (holes in their sides), not through the mouth or anything corresponding to a nose, and the Cambridge group decided that it would be too difficult to attach a mask to collect the air that the bees breathed out. Instead, they repeatedly circulated the same small volume of air through their wind tunnel. The bee removed a little of the oxygen during each circuit of the air, causing the oxygen concentration to decrease gradually (but only by a very little). Indeed, it decreased so little that they had to use a specially modified oxygen analyzer to record the changes. The measurements showed that the bee used oxygen at the same rate at all speeds between hovering and fast flight at 4 meters per second. The rate of consumption was much higher (in proportion to its mass) than for the birds and bats: 300 watts per kilogram. The rough calculation of likely energy cost that we made for birds and bats is inappropriate for bees because their beating wings move through the air many times faster than the body.

The Kori bustard (*Ardeotis kori*) seems to be the largest flying animal: Geoffrey Maloiy and I weighed one in Kenya that had a mass of 16 kilograms (35 pounds). Larger modern birds such as ostriches and Emperor penguins cannot fly, but a few fossil bones have been found of *Argentavis,* a gigantic bird of prey whose mass has been estimated as 80 kilograms (nearly 180 pounds), and which had well-developed wings.

It is difficult for very large animals to produce enough power for flight. We used the rates of loss of height for gliding birds to estimate that a gliding bird requires mechanical power equal to 10 to 25 watts per kilogram. Model gliders lose height at similar rates to full-sized

The Kori bustard (*Ardeotis kori*) of East Africa is probably the largest flying bird. Specimens up to 16 kilograms (35 pounds) have been weighed.

ones, which suggests that the power requirement is the same for aircraft of all sizes. In that case, it would presumably also be the same for small birds as for large ones. However, another argument seems to tell us that larger aircraft need more power per unit mass for flight than smaller ones. Thus flying birds of all sizes must produce *at least* 10 watts per kilogram. To do so, a hummingbird's muscles beating 26 times per second would have to do at least 0.4 joule of work per kilogram of body mass in each beat; a condor's muscles beating 2.6 times per second would have to do 4 joules per kilogram; and the muscles of a larger bird would have to do even more. There must be a size limit above which the muscles cannot do enough work.

The study of flapping flight has presented formidable challenges to research workers. The aerodynamic forces on wings that keep stopping and starting cannot be calculated by the conventional aerodynamics of fixed-wing aircraft, so novel theories have been needed. These relate the forces on wings to the swirling movements of air in the vortices created by the flapping wings, but these vortices could not be photographed satisfactorily and measured until the helium bubble technique had been devised. Similarly, the oxygen consumption of insects in free flight could not be measured until Ellington had perfected his astonishingly sensitive equipment.

Great progress has been made in the past 15 years, but our understanding of flapping flight remains very imperfect. One of the insights that seems most likely to be helpful in future studies of flight is that several gaits are involved. There seems to be a continuous gradation in the flight styles of bees, from hovering to the fastest forward flight, but the slow vortex ring gait of birds and bats seems distinct from their fast continuous vortex gait, and bounding flight is another distinct gait that we will not understand fully until we know more about the physiology of the wing muscles.

# 6

# Buoyancy

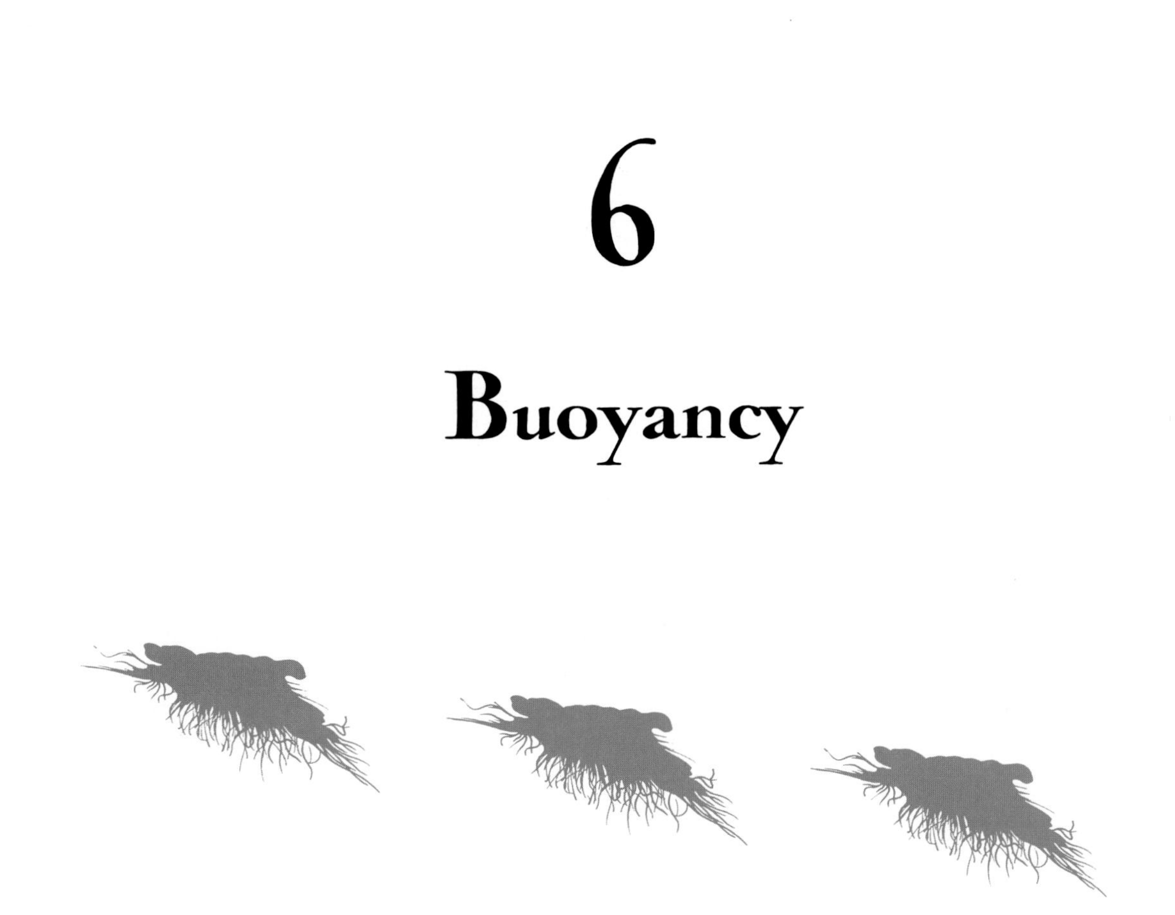

The "jelly" of jellyfishes helps to buoy them up: its chemical composition makes it very slightly less dense than seawater. Although only a single layer of living cells covers the mass of buoyant jelly, many jellyfishes are slightly denser than seawater overall, and sink slowly whenever they stop swimming.

Muscle sinks in water, and so do most of the other materials that animals are made of. These materials sink because they are denser than the surrounding water, whether that water is fresh water, which has a density of 1000 kilograms per cubic meter, or seawater, which has a density of nearly 1030 kilograms per cubic meter. Muscle is denser than either (about 1060 kilograms per cubic meter), mainly because of its protein content. Many of the materials of animal skeletons are reinforced by inorganic crystals and are very considerably denser. Bone consists mainly of protein and calcium phosphate crystals and has a density of about 2000 kilograms per cubic meter. Mollusc shell is calcium carbonate combined with a tiny proportion of protein, and its density is 2700 kilograms per cubic meter.

Unless a water-dwelling animal spends all its time on the sea floor, it must have some way of compensating for its naturally high density. Many swimming animals have floats or other buoyancy organs that reduce their densities to match the water that they live in. You can tell how good the match is, for many fishes, by watching them in aquaria: at times they hang almost motionless in the water, hardly moving their fins, yet they neither sink nor rise to the surface. Other swimming animals are denser than water but have no buoyancy organs. Flounders and most sharks, for example, have densities of 1060 to 1090 kilograms per cubic meter. These animals swim to stay high in the water; as soon as they stop, they sink to the bottom. Although flounders and sharks can be observed resting on the bottoms of aquaria, some fish of similar density, the large and vigorous tunnies, apparently never stop swimming. Squids without buoyancy organs have densities of 1055 to 1075 kilograms per cubic meter and also swim to avoid sinking.

## Swim or Sink

The density of an animal can be measured by a simple procedure that depends solely on being able to weigh the animal, both in and out of the water. A little explanation is needed to show how the procedure works. The weight of an animal is its mass multiplied by the gravitational acceleration, and mass is volume multiplied by density:

$$\begin{aligned}\text{weight} &= \text{mass} \times \text{gravity} \\ &= \text{volume} \times \text{density} \times \text{gravity}\end{aligned}$$

When an animal is submerged in water, an upthrust acts on it, which, according to Archimedes' Principle, equals the weight of a volume of water equal to the volume of the animal:

$$\text{upthrust} = \text{volume} \times \text{water density} \times \text{gravity}$$

Dividing the first equation by the second gives us a formula for density:

$$\begin{aligned}\frac{\text{weight}}{\text{upthrust}} &= \frac{\text{volume} \times \text{density} \times \text{gravity}}{\text{volume} \times \text{water density} \times \text{gravity}} \\ &= \frac{\text{density}}{\text{water density}} \\ \text{density} &= \frac{\text{water density} \times \text{weight}}{\text{upthrust}}\end{aligned}$$

So, to calculate the density of an animal we need to know its weight and the upthrust. To obtain its weight, we can weigh the animal in the ordinary way in air. Since the effect of the upthrust is to reduce the animal's weight in water, we can find the upthrust by weighing the animal again while it is suspended in water and subtracting its weight from its weight in air:

$$\text{upthrust} = \text{weight in air} - \text{weight in water}$$

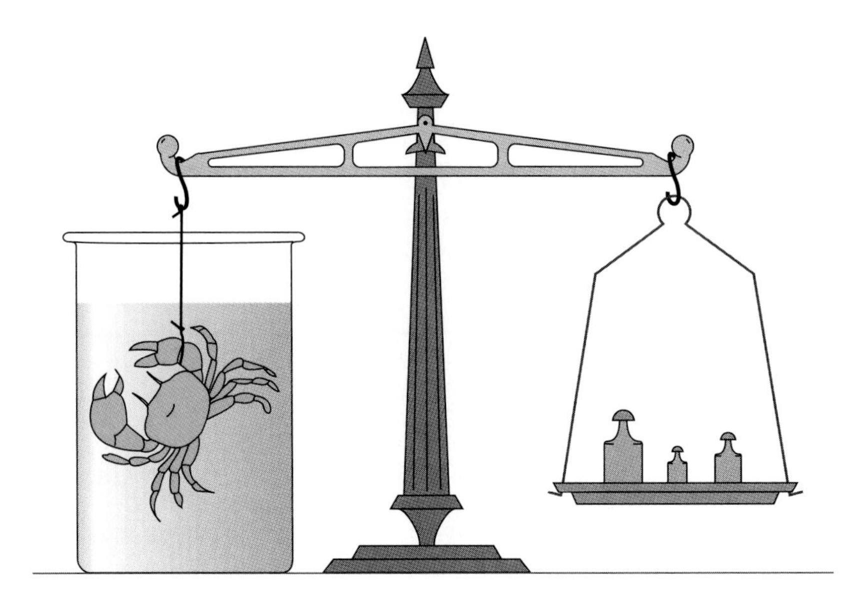

The density of a crab can be measured by weighing the animal in water.

Once we have the two weights, a simple calculation provides the density:

$$\text{density} = \frac{\text{water density} \times \text{weight}}{\text{weight in air} - \text{weight in water}}$$

A simple comparison of the water density and the animal's density will tell us whether an animal will sink or float. Sometimes we want to know not merely whether an animal will sink, but how strong the force is that will make it sink. Just as on land an animal's weight is the force drawing the animal downward, so an animal's weight in water is the force making it sink. We are going to work out how large that force is for animals of particular densities:

$$\text{weight in water} = \text{weight} - \text{upthrust}$$

Using the equations given above for weight and upthrust,

$$\text{weight in water} = \text{volume} \times \text{density of animal} \times \text{gravity} - (\text{volume} \times \text{water density} \times \text{gravity})$$

$$\text{weight in water} = \text{volume} \times \text{excess density} \times \text{gravity}$$

(By excess density I mean the difference between the density of the animal and that of the water.) If we divide each side of the equation by the animal's weight, we get the simpler equation:

$$\frac{\text{weight in water}}{\text{weight}} = \frac{\text{excess density}}{\text{density}}$$

For example, a flounder with a density of 1080 kilograms per cubic meter has an excess density in seawater of 50 kilograms per cubic meter, and its weight in seawater is 50/1080 of its weight in air, or 4.6 percent of that weight. That is reasonably typical of animals that have no buoyancy organs: the sinking force that they have to overcome when they swim in the sea is usually about 5 percent of body weight.

Such animals must produce upward hydrodynamic forces to overcome this sinking force, just as birds must produce lift to fly. Sharks

This shark (*Carcharinus* sp.), photographed in the Red Sea, is denser than water but is preventing itself from sinking by using its pectoral fins like the wings of an airplane.

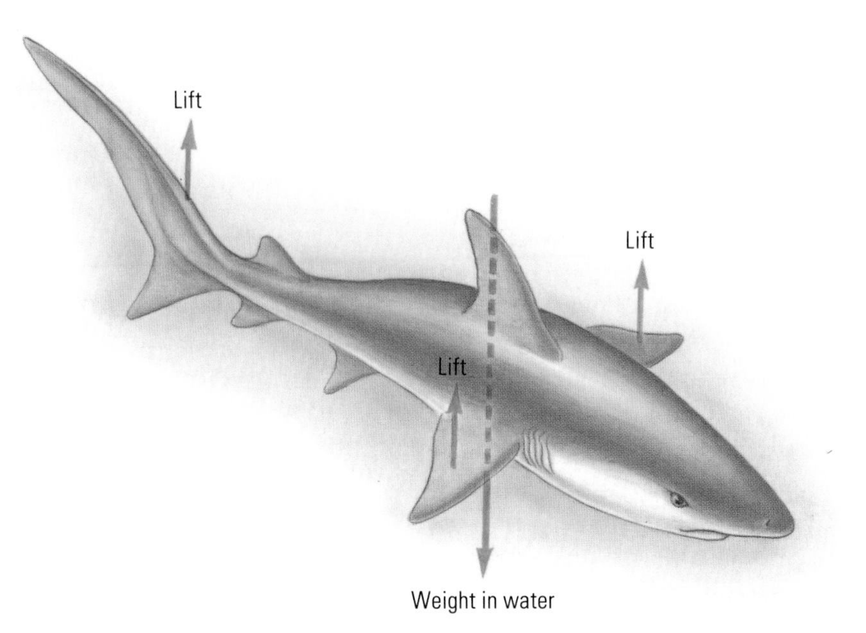


The tail as well as the pectoral fins of a shark must provide lift to counteract the fish's submerged weight.

depend for upward lift on large pectoral fins (the pair of fins, one on each side of the body close behind the head), which are held at an angle of attack. Sharks' tails are asymmetrical, shaped to fan water downward as well as backward when the fish swims, and so provide an upward force as well as thrust. The lift on the fins and the upward force on the tail together balance the animal's weight in water. Because the fins are much closer to the shark's center of gravity than the tail is, most of the upward force has to come from them. Tunnies also depend mainly on their pectoral fins for lift to prevent them from sinking, and squids have fins that function in the same way.

These animals use their fins like airplane wings, to support themselves as they swim along. Some other dense swimmers behave more like helicopters. Copepods are small crustaceans, most of them less than 2 millimeters long, that are very common in the upper waters of seas and are the principal food of herring. They compensate for sinking simply by swimming upward. Some of them hover at fairly constant levels, but others use a "hop and sink" technique, alternately swimming rapidly up and sinking slowly down. That is probably

Fast-swimming squid like these are denser than water.

Copepods are denser than water. They spread their long antennae (as in this picture) to slow their sinking, but fold the antennae out of the way to swim upward. This is *Calanus finmarchicus*, about four millimeters long.

rather an economical way of keeping afloat for these animals because they have long antennae that can be spread to serve as a parachute while sinking but fold away so that the animal is reasonably well streamlined while it is swimming upward.

## Floats from Oily Materials and Salt Solutions

The alternative to lift-producing fins or upward swimming is to have some low-density material in the body that will act as a float. This option has been adopted by a great many swimming animals, which have floats big enough to make the density of the whole animal approximately equal to the density of the water. Some of them have huge volumes of low-density material that swell the body into strange shapes. We will soon see why huge volumes may be needed.

The quantity of low-density material required depends on how low its density is. The weight in water of an animal without a float is, as we have already seen,

$$\text{weight in water without float} = \text{volume} \times \text{excess density} \times \text{gravity}$$

Instead of having a downward weight in water, a float is buoyed with an upward force that can be calculated from its density shortfall (water density minus float density):

$$\text{buoyancy of float} = \text{float volume} \times \text{density shortfall} \times \text{gravity}$$

To give an animal exactly the same density as the water, so that it neither floats nor sinks, the buoyancy of the float must counteract the animal's weight in water:

$$\text{buoyancy of float} = \text{weight in water without float}$$

By replacing the terms in this formula with their equivalents, we can discover how the volume of the float is related to the density of the material:

$$\text{float volume} \times \text{density shortfall} \times \text{gravity} = \text{initial animal volume} \times \text{excess density} \times \text{gravity}$$

$$\frac{\text{float volume}}{\text{initial animal volume}} = \frac{\text{excess density of animal without float}}{\text{density shortfall of float}}$$

If the float has a low density (a large density shortfall), a small float will suffice, but if it is only slightly less dense than the water (having a small density shortfall), it needs to be large.

Ideally, floats should be as small as possible, and therefore of low density. Fats are low-density materials that are quite plentiful in many animals, but they are only a little less dense than water, as can be shown by putting a piece of fat in water: it will float, but with only a very small fraction of its volume above the water surface. Typical fats have densities of about 930 kilograms per cubic meter, giving a density shortfall in seawater of about 100 kilograms per cubic meter. To achieve the same density as seawater, a flounder with an excess density of 50 kilograms per cubic meter would need enough fat to increase its volume by 50 percent. If the flounder lived in fresh water, it would have to be even more obese. Its excess density would then be 80 kilograms per cubic meter and the density shortfall of the fat only 70 kilograms per cubic meter, so the fat needed to make it float would more than double its volume.

Some other oily materials are less dense than fat, and smaller volumes of them suffice to give animals buoyancy. Squalene is a hydrocarbon (that is, a member of the same group of chemical compounds as fuel oils) with a density of only 860 kilograms per cubic meter, found in large quantities in the livers of some sharks. Its density shortfall in seawater is 170 kilograms per cubic meter, so the volume needed to float a fish with an initial excess density of 50 kilograms per cubic meter is only 50/170, or 29 percent, of the initial volume of the fish. To contain even this quantity of squalene, sharks that depend on it for buoyancy have enormously enlarged livers.

The best known of these sharks is the gigantic basking shark, which grows to lengths of at least 11 meters (some books say even longer) and masses of at least 8 tonnes (18,000 pounds). Despite its size, it feeds on small plankton that it strains out of the water as it swims slowly along. Its squalene content makes the shark an attractive target for hunters, who sell the squalene for use in the leather industry.

Wax esters are another group of oily compounds, of about the same density as squalene. They serve to give buoyancy to lantern

Coelacanths were known only as fossils until this species, *Latimeria chalumnae,* was first captured in 1938. It is given buoyancy by wax esters that permeate its tissues.

fishes (described later in this chapter) and also to the coelacanth *Latimeria*. Of all living fish, *Latimeria* is the closest to the evolutionary ancestry of the legged vertebrates, including ourselves. Although the coelacanths are well known as fossils, they were thought to have died out at the same time as the dinosaurs until a Miss Latimer saved one from a South African fisherman's catch in 1938. Several dozen of these large fish, weighing up to 80 kilograms, have been caught since, and I was fortunate to have the opportunity to dissect one. It was a most interesting dissection but a peculiarly unpleasant one because the oily wax esters are not concentrated in the liver, like the squalene of basking sharks, but permeate the whole body.

As an alternative to organic compounds such as squalene and wax esters, animals can gain buoyancy from low-density solutions of salts in water. These solutions cannot be less dense than pure water (density 1000 kilograms per cubic meter) and thus are effective only in salt water, and even there they cannot have large density shortfalls: large volumes of these solutions are needed to compensate for even small excess densities in other parts of the animal. The jelly of jellyfish is slightly less dense than seawater because it contains a smaller proportion of heavy sulfate ions. It is a very dilute protein gel containing only about 1 percent organic matter. (In contrast, table jelly needs 5 percent gelatin to be reasonably rigid.) Jellyfish jelly also contains dissolved salts in the same osmotic concentration as the salts in seawater: if the jelly were more dilute, its water would be drawn out by osmosis and the jellyfish would shrivel up. Its density is slightly reduced because it contains less sulfate than seawater does, and so contains correspondingly more of other, lighter ions, but the

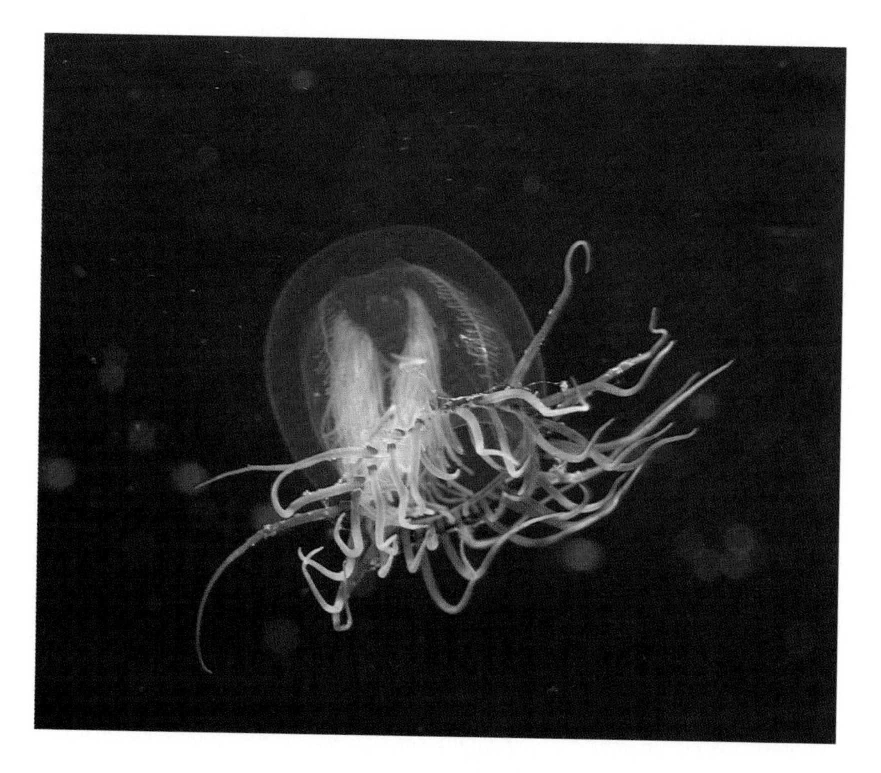

A small jellyfish (*Polyorchis montereyensis*, about 4 centimeters or $1\frac{1}{2}$ inches in diameter). This species is slightly denser than water, so it sinks whenever it stops swimming.

effect is small because only 8 percent of the salt in seawater is sulfate. The density shortfall of jellyfish jelly in seawater is only about 1 kilogram per cubic meter; in consequence, a huge volume of jelly is needed to buoy up a tiny volume of living tissue. Indeed, the living tissues of jellyfish are confined to an extremely thin layer only one cell thick on the surface of the nonliving jelly. Some jellyfishes are almost exactly the same density as the water they live in, but others are denser and must swim to keep afloat.

Solutions of salt water bring the densities of many squids very close to that of seawater. In these squids, the body cavities or some of the tissues are swollen by large volumes of a solution that resembles seawater but for the fact that most of the sodium has been replaced by ammonium ions. These solutions have densities of about 1010 kilograms per cubic meter, so their density shortfall in seawater is almost 20 kilograms per cubic meter—much more than the density shortfall of jellyfish jelly but much less than that of fats, squalene, and wax esters. Some squids look strangely bloated because they must hold large volumes of these solutions to achieve the same density as the water they live in.

These bloated squids live in the oceans, many of them at depths of several hundred meters. Although they are seldom seen alive, their remains are found more commonly than any other food in the stomachs of sperm whales. They must be very numerous, for it has been calculated that the quantity eaten annually by sperm whales exceeds the annual fish catch of the combined fishing fleets of the world.

## Chambers of Gas

Floats filled with gas can be much smaller than other kinds because gases have very low densities. For example, at room temperature and atmospheric pressure, the density of air is only 1.3 kilograms per cubic meter, about $\frac{1}{800}$ the density of water.

One ingenious method for bringing gas into a float is used by the pearly nautilus, a primitive relative of the squids. It is found in coastal waters in the southwest Pacific, where it can be caught in lobster pots. Though related to the squids, it has a large shell, coiled

The pearly nautilus has gas-filled buoyancy chambers in its shell.

like the shells of some snails. The soft parts of the animal's body occupy the quarter-turn of the shell nearest the opening, and the rest of the shell is divided into a series of gas-filled chambers. Even though the gas has a very low density, quite a large shell is needed to make the animal float, because the shell itself is so dense.

Nautiluses are caught at considerable depths, often at 100 meters or more. Hydrostatic pressure in the sea increases by 1 atmosphere for every 10 meters of depth, so nautiluses caught at 100 meters must have been living at high pressures. You might imagine that the gases in the shell chambers would also be at high pressures and that these gases would bubble out if a freshly caught animal was held under water while a hole was bored into its shell. Eric Denton and J. B. Gilpin-Brown (then based in the marine laboratory in Plymouth, England) tried the experiment and got the opposite result: gas did not bubble out but water was sucked in, showing that there had been a partial vacuum in the intact shell. The two experimenters measured the volume of water sucked in by weighing the nautilus before and after the experiment, and with this information they were able to calculate the gas pressure: it was about 0.8 atmosphere in the older chambers, nearest the center of the spiral shell, and as little as 0.3 atmosphere in some of the newer chambers. The gas was mainly nitrogen.

Besides containing this low-pressure gas, the chambers contain a small amount of a solution of salts in water. Denton and Gilpin-Brown measured the freezing point of this solution (strong solutions freeze at lower temperatures than weak ones) and showed that the solution was more dilute than the animal's blood. Strong solutions can suck water from weak ones, if the two solutions are separated by a membrane that allows water but not the dissolved salts to pass through. The process is called osmosis, and its effect is to make the strong solution weaker and the weak one stronger, so that their concentrations become more nearly equal. Water could be drawn by osmosis out of the dilute solution in the chambers and into the blood, but only under certain conditions of pressure. Osmosis can be prevented by a pressure difference sucking in the opposite direction—if the weak solution is at a lower pressure than the strong one. The pressure difference needed to prevent osmotic flow is called the difference in osmotic pressure between the solutions. Denton and Gilpin-Brown's measurements showed that the difference in osmotic pressure between nautilus blood and chamber fluid was 15 atmospheres. That means that water could be drawn out of the chambers by osmosis into the blood against pressure differences of up to 15 atmospheres.

These observations seem to show that the nautilus does not fill its shell chambers by pumping gas into them; rather, it sucks water out osmotically. The pressure in the sea is 1 atmosphere at the surface and increases by 1 atmosphere for every 10 meters of depth, so it is 15 atmospheres at a depth of 140 meters. Thus the observed osmotic pressure difference is just enough to suck water out of the chambers at that depth, leaving vacuum behind. After the chambers have been emptied in this way, gases will diffuse in from the animal's tissues. There is less gas in the newer chambers because there has been less time for gas to diffuse in.

The dissolved gases in the water at the surface of the sea are in equilibrium with the air, at a pressure of 1 atmosphere. Water deeper in the sea contains less dissolved gas, because the animals living in it use up oxygen. As it passes through the gills, the animals' blood

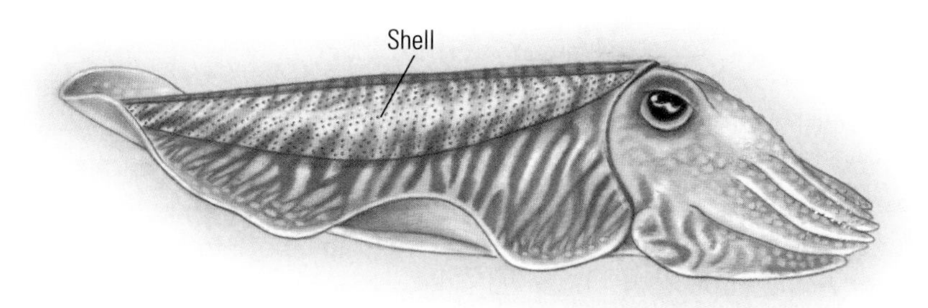


Cuttlefish have a lightly built internal shell that contains a stack of gas-filled chambers.

exchanges gases with the water. Consequently, although these animals are living at depths where the pressure is many atmospheres, their blood and other tissues contain no more gas than would dissolve in them if they were kept in contact with air at 1 atmosphere pressure. That is why the pressure of the gas in the shell of a nautilus never rises above 1 atmosphere, however deep in the sea it is living.

Cuttlefish, other relatives of squid, have a more buoyant shell than the nautilus. This is the cuttlebone, which is often given to caged birds to peck. It has no protective function but is buried inside the body. It consists of gas spaces sandwiched between thin layers of shell, and its very light construction gives it a density of only 620 kilograms per cubic meter, less than the density of any of the buoyancy aids discussed so far. It occupies only 10 percent of the volume of the cuttlefish, but that is enough to make the density of the animal almost exactly the same as that of water. Cuttlefish can hang almost motionless in midwater or move slowly around by undulating their fins.

The rigid gas chambers of the nautilus and cuttlefish, filled with gas at low pressure, would collapse if the animals swam too deep. Denton and his colleagues put the shells of these animals into pressure chambers and increased the pressure until they collapsed. Nautilus shells collapsed at about 65 atmospheres (the pressure at a depth of 640 meters), but nautiluses seem not to swim deeper than about 500 meters. Cuttlebones collapsed at 24 atmospheres (the pressure at 230 meters), but cuttlefish seem to live no deeper than about 150 meters. These animals do not seem to go deep enough to put their shells at risk.

## The Swimbladder

The most familiar fishes (salmon, herring, cod, and a great many others) belong to the group called the teleosts. Many of them have a gas-filled float, but unlike the floats of the nautilus and cuttlefish, this float has walls that are flexible enough for there to be no danger of its being broken by high outside pressures. This float is the swimbladder, a bag of gas in the body cavity close under the backbone. Its walls are thin and little denser than water, so we can ignore them in calculating the volume a swimbladder should have. The swimbladder is even more buoyant than the cuttlebone, having a density shortfall in seawater of about 1000 kilograms per cubic meter. To correct the typical flounder's excess density of 50 kilograms per cubic meter, for example, all that would be needed is a swimbladder that added 5 percent to the initial volume of the fish. Swimbladders are much smaller than the other buoyancy aids we have discussed, because they are so much less dense. This advantage of swimbladders must be weighed against serious disadvantages.

Several disadvantages arise from the flexibility of the swimbladder wall. One is that gas leaks from the swimbladder. Its flexible walls allow the structure to shrink and the gases in the swimbladder to be compressed to the pressure of the surrounding water, which is many times greater than atmospheric pressure if the fish is at a substantial depth. However, there is no more gas in the fish's tissues than there would be if the gas was in equilibrium with air at atmospheric pressure, for reasons that have already been explained. Because the pressure of the swimbladder gases in a submerged fish is always greater than the pressure at which they would be in equilibrium with the tissues, gases must diffuse out of the swimbladder unless its wall is utterly impermeable to them, which it cannot be. If the fish is to remain buoyant, the slowly leaking gases must be replaced. The problem is much less severe than it could be, however. George Lapennas and Knut Schmidt-Nielsen of Duke University have shown that swimbladder walls are far less permeable to gases than other tissues are. This is because they have embedded in them layer upon layer of ribbonlike crystals of a compound called guanine, so thin that they are flexible. Diffusion through the crystals is exceedingly slow because the gases have to take a tortuous route around

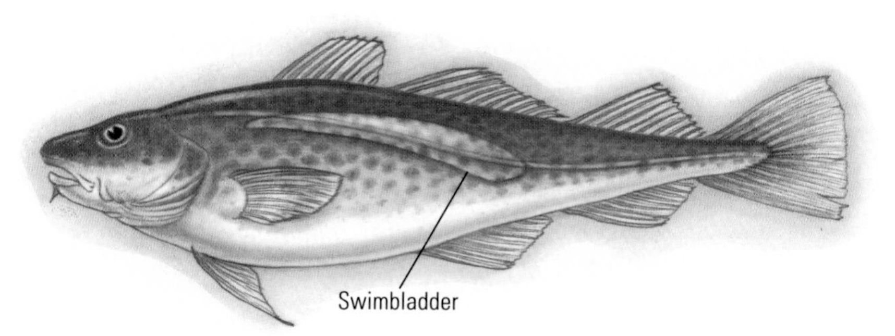


Most teleost fishes have a gas-filled swimbladder that gives them buoyancy.

and between the impermeable crystals. Guanine crystals are also present under the scales of many fishes and give them their silvery sheen.

Another disadvantage of the flexible swimbladder wall is that it allows the volume of the bladder to change as the fish changes depth, altering the buoyancy of the fish. As the fish swims deeper, the increasing pressure compresses the swimbladder, making it smaller. As the fish swims up to shallower depths, the gases expand as the pressure falls and the swimbladder becomes larger. Unless the quantity of gas in the swimbladder is changed, the swimbladder can have the size needed to make the fish the same density as the water only at one particular depth. Imagine a typical fish that, in order to match its density to that of the water, needs a swimbladder that is 5 percent of the volume of the rest of its body. We will suppose that its swimbladder has that volume at the surface, where the pressure is 1 atmosphere. As the fish swims 10 meters down to where the pressure is 2 atmospheres, the swimbladder is compressed to 2.5 percent of the volume of the rest of the body, leaving the fish with 2.5 percent excess density. If the fish swims on down to 90 meters where the pressure is 10 atmospheres, the gas is compressed to one tenth of its volume at the surface and is almost useless as a buoyancy aid.

The situation is worse for a fish that starts at 90 meters, with enough gas in its swimbladder to match its density there to that of the water, and ascends almost to the surface. As the fish rises, the swimbladder expands from its initial 5 percent of the volume of the rest of the body to 50 percent. The fish becomes much less dense than the water and must swim strongly downward if it wants to avoid being carried right to the surface. Even worse, the swollen swimbladder might burst or severely injure the fish. Fish such as hake that are caught on lines at depths close to the 90 meters of our example arrive at the surface so hugely bloated by their expanding swimbladders that the stomach is often forced out of the mouth, turned inside out. For such fish, a quick trip to the surface is disastrous.

If a fish with a swimbladder moves even slightly up from the depth at which its density is matched to that of the water, it will become less dense and continue to rise unless it makes a correcting movement. If it moves slightly down, it will become denser and sink. A fish with a swimbladder adjusted to its depth is in unstable equilibrium. Similarly, tightrope walkers are in unstable equilibrium because if they lean to the left they will fall farther to the left (unless they make a correcting movement) and if they lean to the right they will fall farther to the right. Even if a fish of perfectly adjusted density remains motionless, some chance movement in the water will soon make it start rising or sinking. Once started, it will continue to rise or sink. If you watch fish hanging in midwater in aquaria, you will seldom see their fins remain motionless for long, because small correcting movements are constantly needed. Similarly, tightrope walkers must make small correcting movements constantly: it is dangerous to fall asleep on a high wire.

If fish are to be able to adjust their swimbladders to suit different depths, they must be able to add gas when they go deeper and remove it when they approach the surface. Even if they were to spend their entire lives at one depth, they would have to add gas to the swimbladder to replace the gas that inevitably diffuses away.

Some fish have a tube leading from the swimbladder to the mouth and can swallow air or spit it out to adjust the quantity in the swimbladder. A fish at any depth can spit air out, but a fish that is already at a substantial depth cannot gulp air in to adjust its density as it swims to greater depths. Fish need some means of adding gas to the swimbladder other than simply gulping it in.

The connection to the mouth seems to have been present when the swimbladder first evolved, but most fish with swimbladders have lost it, which is a little surprising as it seems to be a useful safety valve. A fish that has risen too high in the water and is in danger of being burst by its expanding swimbladder can save itself quickly by burping, if it has this connection to the mouth.

The arrangements for adjusting the quantity of gas in the swimbladder differ between groups of fish, but I will describe a common arrangement found, for example, in cod (which has no connection from swimbladder to mouth). In such fish, gas is brought to the swimbladder and carried away through the blood. Most of the wall of a cod's swimbladder has a very sparse blood supply, but there are quite large arteries leading to two regions, the oval and the gas gland.

The oval is a pocket in the swimbladder wall that is usually kept closed off from the rest of the swimbladder by a ring of muscle around its mouth, which closes it like a drawstring purse. The blood vessels leading to the oval are usually kept constricted to stop the flow of blood. When there is a need to remove gas from the swimbladder, a hormone is released into the blood that makes the drawstring muscle relax, allowing the oval to open, and that makes the blood vessels dilate, letting blood flow through the wall of the oval. As it passes through the oval wall, the blood comes into close proximity to the swimbladder gases. We have already seen that, given the opportunity, gases will diffuse out of the swimbladder into the blood. The opening of the oval and the dilation of its blood vessels allow gases to diffuse quite rapidly out of the swimbladder.

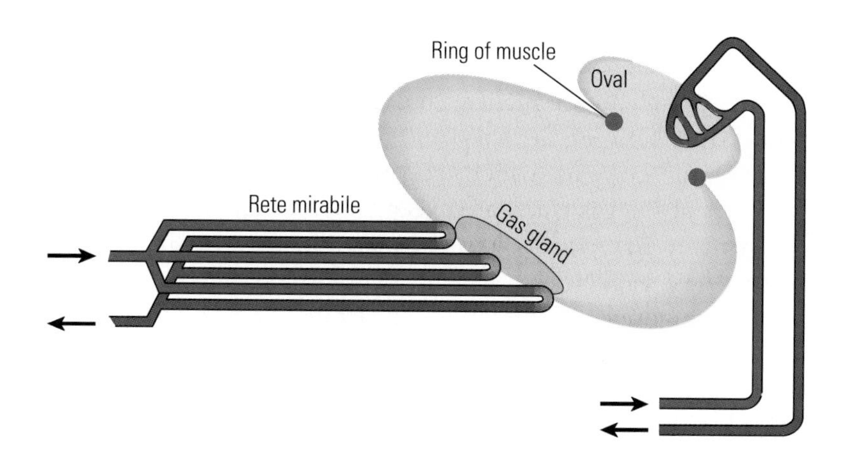


A typical swimbladder and its blood supply. Gas is secreted into the swimbladder by the gas gland or resorbed by the oval, as required.

Removing gas from the swimbladder is simply a matter of allowing it to escape; rather similarly, to remove air from bicycle tires you have only to open the valves. In contrast, adding gas to the swimbladder involves forcing it in. It is equivalent to inflating a bicycle tire, which requires a pump to compress the air.

The swimbladder's "pump" is located at the gas gland, a patch on the swimbladder wall that is red because it has a plentiful blood supply. It does not work in the same way as a bicycle pump, but it performs an equivalent task. When it needs to release gases, it adds lactic acid to the blood flowing through it. (This is the chemical substance that is also produced in muscles during anaerobic metabolism, when an oxygen debt is building up.) The effect of the lactic acid is to release gases from the blood. Oxygen is carried, from the gills to the parts of the body where it is needed, in combination with hemoglobin, the pigment that gives blood its red color; lactic acid releases some of the oxygen from the hemoglobin. It also releases carbon dioxide from bicarbonates in the blood and makes all gases slightly less soluble in the blood. These effects would be sufficient to release gases into the swimbladder in fish living near the surface of the water, where the pressure of the swimbladder gases is little more than atmospheric pressure. By themselves these effects are far too weak to force gases into the high-pressure swimbladders of deep-water fish.

The effect of the lactic acid is greatly amplified by a structure called the rete mirabile. (These Latin words mean "wonderful network.") When arteries reach the tissues that they serve, they break up into extremely fine branches called capillaries, which are commonly about half a millimeter long. The whole tissue is permeated by this network of fine branches, so oxygen and other substances can easily diffuse between the blood and the tissue. The capillaries merge again to form the veins, which carry the blood back to the heart. The capillaries of the gas gland are exceptionally long, about 10 millimeters in many fish and as long as 25 millimeters in some deep-sea fishes. These capillaries constitute the rete mirabile. In that structure, the hundreds of capillaries run parallel to each other as far as the gas gland and then loop back in such a way that the arterial parts of the capillaries (leading to the gland) run alongside and are intermingled with the venous parts (leading back from the gland). Lactic

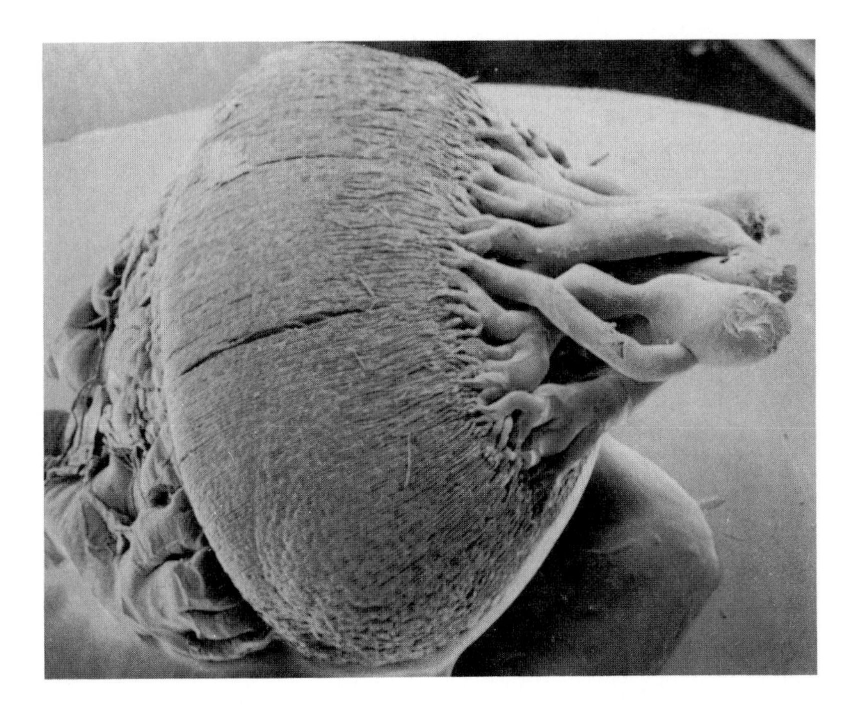

The rete mirabile of an eel (*Anguilla anguilla*) swimbladder. The arteries and veins divide into a huge number of fine capillaries which run parallel to each other for a distance of several millimeters. This preparation was made by injecting latex into the blood vessels. After it had set, the tissues were dissolved away leaving only the latex, which is seen here by scanning electron microscopy.

acid is added at the turn of the loop so that the returning blood in the venous capillaries contains more lactic acid than the blood arriving in the arterial capillaries. Gases are released from the blood in the venous capillaries by the lactic acid and diffuse across to the arterial ones; consequently, the blood arriving at the gas gland now contains increased quantities of gases. When lactic acid is added in the gland to this gas-enriched blood, more gas is released from it—and this gas diffuses across from the venous to the arterial capillaries, increasing still further the quantity of gas in the blood arriving at the gland. The effect is that gases become trapped in the looped capillaries, diffusing across from the venous to the arterial capillaries instead of continuing on their circuit around the body. Gases accumulate until the effective pressure of the gases in the blood passing through the gas gland is greater than the pressure of the gases in the swimbladder. When this happens, gases diffuse from the blood into the swimbladder. The mechanism is effective enough to fill the swimbladders of fish living at depths of a few kilometers, where the pressure is a few hundred atmospheres. (For comparison, bicycle tires are inflated to pressures of 10 atmospheres or less.)

The rates at which gases can be secreted into swimbladders and removed from them were first measured more than a hundred years ago by a French scientist, Armand Moreau, who put wrasses (*Labrus bergylta*) into cages and suspended them at various depths from the end of a pier. He transferred fish that had been living in shallow water to depths of up to 8 meters and hauled them up to the surface from time to time to find out how much gas had been secreted into the swimbladder.

A much more sophisticated version of the same experiment was performed a hundred years later by Roy Harden Jones and P. Scholes at the fisheries laboratory at Lowestoft, England. They studied the adjustment of swimbladders to varying pressure by observing cod kept in a pressure tank with a viewing window. To simulate changes in depth, they raised the pressure in the chamber by means of a pump that could achieve a maximum pressure of 7.5 atmospheres, the pressure in the sea at a depth of 65 meters. When fish accustomed to low pressures were suddenly subjected to high ones, they were seen at first to be quite obviously denser than the water—they rested on the bottom or had to swim quite vigorously to get off it. Only after many hours were the fish able to recover their buoyancy. During the period of recovery, their progress was checked occasionally by briefly reducing the pressure by various amounts and observing the pressure at which they just floated. Similar experiments were used to measure rates of removal of gas, after a reduction from an initially high pressure. When the pressure was increased to simulate movement to a greater depth, the cod compensated for the apparent depth change at a rate of only 1 meter per hour. When the pressure was decreased, simulating a reduction in depth, they compensated at rates that depended on the new depth. At 5 meters they compensated for depth reduction at only 1 meter per hour, whereas at 65 meters they compensated at 20 meters per hour. At the greater depth there is more pressure in the swimbladder to drive the diffusion of gases to the blood vessels of the oval.

Some other fish can fill their swimbladders faster than cod (for example, bluefish can secrete gas fast enough to compensate for depth changes at 2.5 meters per hour), but none seems able to secrete gas fast enough to compensate for fast descents. Still, some fish that have swimbladders regularly change their depth very rapidly indeed. The surprising ascents and descents of these fish were shown dramatically by echo sounding, the technique of measuring the depth of the sea by bouncing sound waves off the bottom and measuring the time required for the echo to return. Sound travels in water at 1400 meters per second, so if an echo returns (for example) after an interval of 1 second, it has traveled a total distance down and up of 1400 meters, and the water must be 700 meters deep.

When sound waves are directed toward the bottom of the ocean, the results are a bit surprising: there is not only the sharply defined echo from the bottom, but generally also a fuzzy echo from a "deep scattering layer." This layer is near the surface at night but descends to a depth of several hundred meters at dawn, where it remains all day until returning again to near the surface at dusk. The U.S. Navy became interested, presumably because echo sounding is used for detecting enemy submarines as well as for measuring the depth of the sea bottom, and submarines would be less conspicuous to echo sounding if they were stationed in the deep scattering layer. The Navy sponsored research and an international conference on the deep scattering layer.

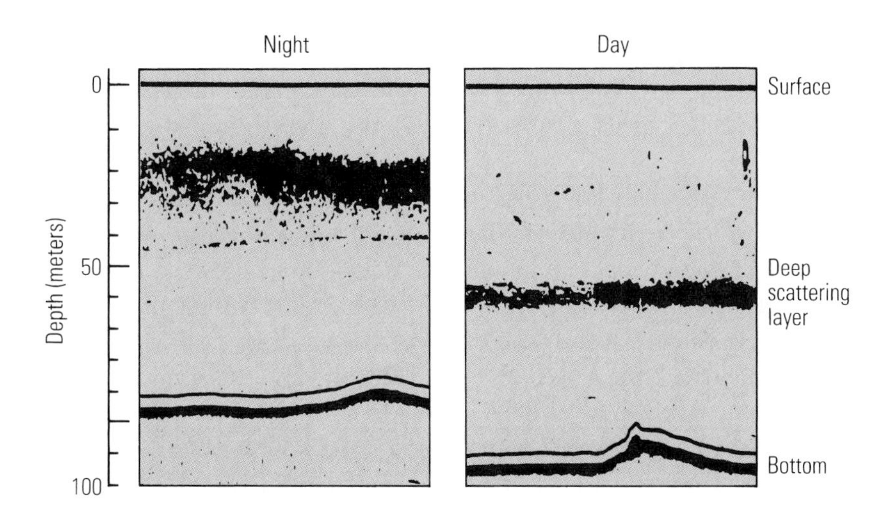


Echo sounders record echoes from the sea bottom and also from animals such as the fish and siphonophores of the deep scattering layer. Black bands show the depths from which echoes were received over a period of time during which the ship that made the record was moving.

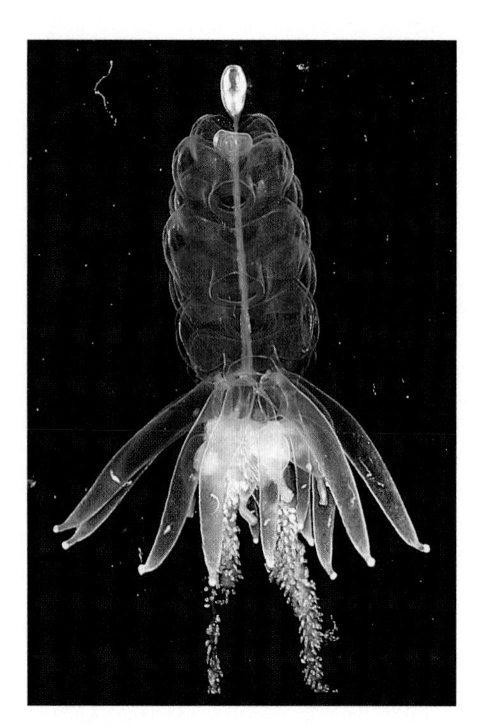

This siphonophore (*Physophora hydrostatica*) has a small gas-filled float at the top. Below that is a cluster of swimming bells that pulsate, drawing in water and squirting it out to power swimming. At the bottom are tentacles with stinging cells (nematocytes) that are used to capture prey.

Scientific fishing with special nets designed to catch fish at particular depths gave a strong indication that the deep scattering layer might consist of small fishes, shrimps, and siphonophores—simple animals that resemble a chain of linked jellyfishes. These animals were caught near the surface at night and at the depth of the deep scattering layer by day. The most plentiful of the fishes were lantern fishes, sardine-sized fish that have rows of light-emitting organs along the underside of the body. Some lantern fishes, but not all, have swimbladders, and both types were caught at the two depths. Clearer evidence that the deep scattering layer contained lantern fishes came when Eric Barham, a scientist at a U.S. Navy laboratory, observed the layer from a deep submersible vehicle—a research submarine capable of diving to substantial depths. By day, all the lantern fishes seen in dives in the Pacific Ocean off California were at depths of 250 meters or more, but by night a lot of them had moved within 50 meters of the surface. Some other fishes had moved up only a little by night or had stayed at about the same depth, but many of the lantern fishes made daily ascents and descents of about 300 meters. By following the fish in his submersible, Barham confirmed that they really did make these very large vertical movements. Ships at the surface used echolocation while the submersible dived to confirm that the submersible was indeed in the deep scattering layer while the fish were being observed.

A cod, secreting gas at the rate measured by Harden Jones, would take 300 hours to compensate for a 300-meter depth change. A bluefish, secreting at its higher rate, would take 120 hours. Lantern fishes complete their morning descent in only about 3 hours. It seems inconceivable that they can secrete gas fast enough to maintain their buoyancy.

Only some lantern fishes have gas-filled swimbladders. When these have been caught near the surface at night, their swimbladders have been found to contain about the right volume of gas to make the fish the same density as the water. It would be very difficult to measure their density at daytime depths, but we must assume that the swimbladder is compressed to a tiny fraction of its nighttime volume, making the fish almost as dense as if they had no swimbladder at all. They must have to keep swimming to avoid sinking further.

Other lantern fishes have gas-filled swimbladders when they are young, but lose the gas as they grow older, when wax esters accumu-

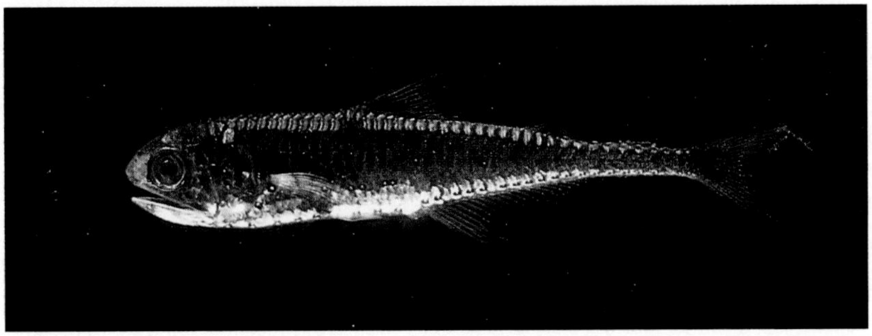

Lantern fishes (family Myctophidae) like this one are extremely common in the oceans, where they spend the night near the surface and the day at depths of several hundred meters, where only dim light penetrates. These fish spend their lives in near-darkness but have light-emitting organs, visible here as bright spots on the belly.

late around the remnant of the swimbladder and in their muscles. The buoyancy of the wax esters replaces the buoyancy of the lost swimbladder. The densities of these fish are about the same as the density of the water and must remain almost precisely constant as the fish swim up and down, which may give them an advantage. Some of them have been observed hanging motionless in the water at depths of around 300 meters, at which depth a gas-filled swimbladder would be compressed very small. (Puzzlingly, the same observation has also been made of lantern fishes that were identified as belonging to a species with a swimbladder. Perhaps the identification was wrong.)

## The Best Buoyancy Policy

This chapter has shown the variety of ways that aquatic animals stay afloat. Some aquatic animals are denser than the water they live in; they have to swim upward to keep off the bottom or rely on fins for hydrodynamic lift. The others have their densities reduced to match that of the water by floats that may be low-density organic compounds, solutions of unusual ionic composition, or gas-filled chambers. None of these possibilities is necessarily superior to the others. Each has its merits, and the best buoyancy policy for a particular animal depends on where it lives and what its habits are.

Fish and other animals that live in the upper waters of deep oceans do not have the option of resting on the bottom, because it is too far down. For those that live in shallower water, the bottom may be the best place to hide or to lie in ambush for prey, because it is easier to hide on the bottom than while floating in midwater. Many swimming animals such as dogfish and flounders have evolved camouflage patterns that make them very inconspicuous when resting in suitable places on the bottom. Many midwater fish such as herring have silvery sides, which make them reasonably inconspicuous in open water, and a few midwater fish are largely transparent, but it remains true that the bottom is the easiest place to hide. For the animals that rest there, being denser than water may be an advantage because the animals will rest firmly on the bottom and not be wafted along by the slightest current. The penalty for being denser than water is that they cannot rest in midwater but must keep swimming, so long as they want to stay off the bottom.

Bottom-living fish such as dogfish and flounders are not the only ones that are denser than water. Porbeagle sharks that swim perpetually are almost as dense as dogfish and, like them, depend on lift from fins. Tunnies seem never to rest on the bottom (at least in aquaria),

Dogfish such as this *Scyliorhinus canicula* (a European species) are denser than the seawater that they live in and spend most of their time resting on the bottom.

yet they are as dense as flounders and also have no swimbladder. Many squids that are denser than water swim perpetually using fins and their jet-propulsion style of swimming to keep them from sinking. Can there be an advantage for animals like these in being denser than water?

Whether hydrodynamic lift or buoyancy organs are used, there is a cost to be paid for keeping off the bottom. To get hydrodynamic lift, work must be done against induced drag, as was explained in the context of flight in Chapter 4. Buoyancy organs do not increase induced drag, but by making the body larger they increase the drag that acts on it in swimming. The options available to swimming animals as they evolve are to have a small, dense body and do work against induced drag, or to have a buoyancy organ that makes the body larger (for the same volume of living tissue) and do more work against drag on the body. The induced drag (for the same lift) decreases as speed increases, but the extra drag on a larger body increases as speed increases. This suggests that slow swimmers will do best with buoyancy organs and that fast swimmers will use less energy if they have no buoyancy organ, but get lift from fins.

Swimming animals seem to have evolved accordingly. The dense squids swim around quite fast, whereas cuttlefish and the nautilus (which have buoyant shells) are relatively sluggish. The sharks that hunt for large prey are dense, but the basking shark, which swims very slowly straining out plankton, is buoyed up by squalene. Although most teleost fish have swimbladders, the fast-swimming tunnies have none.

The advantage changes from buoyancy organs to hydrodynamic lift at a particular swimming speed, and that speed must depend on the composition of the buoyancy organs. The buoyancy organs of lowest density can be smallest, so they increase drag least. In this respect, gas chambers are better than squalene and wax esters, which in turn are better than low-density solutions. Other considerations, however, may change this order of preference: we have seen how wax esters may be better than a swimbladder for lantern fishes that make large daily depth changes. The speed at which the advantage shifts from having a buoyancy organ to doing without one must be slow for the very bulky low-density solutions, faster for moderately bulky squalene, and faster still for compact swimbladders. I have calculated that for 1-kilogram fish this critical speed should be about 0.5 meter per second for squalene and 0.8 meter per second for swimbladders; for 1-tonne fish it should be 3 and 4 meters per second, respectively. Although rather few data are available, it seems doubtful whether tunnies swim quite fast enough for the loss of the swimbladder to give a clear energy-saving advantage. Thus other possible advantages have been suggested instead. For example, the killer whales that feed on tunnies emit sounds and seem to use the echo to locate prey. Air-filled cavities such as swimbladders return strong echoes and would make their possessors conspicuous to echolocation.

Although most animals that live below the water's surface are as dense as or denser than water, a few occasional visitors have the opposite trait. These are the diving birds, which face the problem of being too buoyant.

Birds carry a good deal of air in their bodies, in their lungs and associated air sacs. They also have air trapped between their feathers and skin that has an important heat-insulating function, like the air trapped around human bodies by clothing. All this air gives birds low

This Blue-footed booby (*Sula nebouxii*) has dived into the sea and is attacking a shoal of fish from below.

densities, which is why they ride so high when floating on water. When ducks dive, they have to swim strongly downward to overcome the buoyancy of the air that they carry with them.

The conclusion from this chapter must be that there is no one best buoyancy policy. It may be best for bottom-living fish to be denser than water, so that they sit firmly on the bottom. For fast-swimming fish and squids it may be better to remain a little denser than the water than to be swollen by a large buoyancy organ; but slow-swimming fish and cuttlefish may do better with buoyancy organs that match their density to that of the water. For fish that remain at fairly constant depths, the swimbladder is likely to be the best type of buoyancy organ because it can be the least bulky, but because swimbladders are compressed by increased pressures, wax esters may be better for fish that make large vertical migrations.

# 7

# Swimming

Penguins use their tiny wings to "fly" underwater, but they often leap through the air as in this picture. One theory is that leaping saves energy by enabling the animal to escape from the drag forces that resist its movement through water; another is simply that the bird leaps to breathe. These are Adelie penguins (*Pygoscelis adeliae*).

Now that we have seen how animals stay afloat, we can look at how they propel themselves through water or over its surface. All animals apply the same principle in order to move in water: they push water backward to drive themselves forward. Yet they do that in several very different ways: they may writhe like snakes, shoot like rockets, or beat their tails.

## Living Rowboats

Some animals row themselves along like tiny boats. In rowing, the blade of an oar is pushed backward through the water during the power stroke. Because the oar is moving backward, the drag on it acts forward, serving as thrust to propel the boat. As the boat moves forward, drag acts backward on it: the forward thrust from the oars must balance the backward drag on the boat. Water beetles swim like this, using their legs as oars.

To work efficiently, an oar needs a large blade. The same thrust can be obtained by pushing a lot of water backward at a low speed or a little water backward at a higher speed. Less power is needed if the oars push a lot of water at low speed, and for that the oars must be large. The same principle applies to flight, as we saw in Chapter 4: longer wings give the same lift with less induced drag because they push on more air.

The legs of water beetles have evolved to be good oars. The segments of each leg are flattened to form the central part of the oar blade, which is greatly enlarged by a fringe of long bristles. Because these bristles are closely spaced, they push on the water almost as effectively as would a blade made of a continuous sheet of material of the same length and breadth. A big advantage of the bristles is that they can be folded when required, making the blade much narrower.

After each power stroke, an oar must be brought forward again in a recovery stroke, ready for the next power stroke. During the recovery stroke, any drag on the oar must act backward and slow the animal down. A water beetle (or a person rowing a boat) wants as much drag as possible to act on the oar in the power stroke and as little as possible in the recovery stroke. People rowing boats lift the oars clear of the water for the recovery stroke, but animals do not do that, nor can they, if they are swimming far below the surface. Water beetles spread their leg bristles for the power stroke and fold them for the recovery stroke, when they also reduce drag by turning their flattened leg segments edge on to the water. The spreading and folding are done automatically by the pressure of the water on the bristles, which are hinged to be free to fold one way but not the other.

Ducks paddle with their feet. The principle is the same as rowing, but the feet move alternately like the paddles of a canoe instead of together like the oars of a rowboat. Like the legs of water beetles, the webbed feet are spread for the power stroke and folded for the recovery stroke. The feet paddle in the same way whether the bird is swimming on the surface or diving under water.

A water beetle (*Tropisternus*) diving. The bristles on its legs are spread for the power stroke of rowing but fold out of the way for the recovery stroke.

## Swimming with Wings

Penguins also swim, but their method of swimming is strikingly different from that of ducks and (as we will see) much more economical of energy. Rather than using their feet, penguins swim with their wings and use them as hydrofoils rather than as paddles. The difference is that the thrust from paddles such as ducks' feet is drag, whereas the thrust from hydrofoils such as penguins' wings is lift. The blades of the propellers of ships are also hydrofoils (and those of airplanes are airfoils) that provide lift to serve as thrust. The penguin's method of swimming is also used by sea turtles, which beat their flippers up and down to propel themselves by lift.

A Galapagos penguin (*Spheniscus mendiculus*) fishing in a school of anchovies. It is using its wings to "fly" through the water.

Although the wings of penguins are shaped more or less like those of other birds, they are too small to fly with. The area of one wing of, say, a 4.2-kilogram penguin is 74 square centimeters, whereas that of an eagle of the same body mass would be 2600 square centimeters. The action of the wings as they beat up and down in the water does indeed look very much like that of flying birds, but there is an important difference. The downstroke is like the downstroke of flight: the wing has a positive angle of attack and provides upward, forward lift. The upstroke, however, is quite different. In flight, any lift during the upstroke acts upward and backward, but in the upstroke of penguin swimming the wing is given a negative angle of attack so that the lift acts downward and forward. The downstroke provides an upward, forward force and the upstroke a downward, forward one. The upward and downward components cancel out and the overall effect is the forward thrust that the penguin needs to propel it through the water. The reason for the difference between flight and the swimming of penguins is that the principal component of force required for flight is vertical, to counteract gravity, whereas the principal component of force that penguins need for swimming is horizontal. Penguins do not need a vertical force because they are about the same density as the water. This interpretation of penguin swimming is based mainly on films of captive penguins that had white lines painted on their wings, from the front edge to the rear, to make it easier to measure the wing's angle of attack.

When penguins and ducks swim at the surface of the water, they inevitably push a wave along in front of them, like the bow wave of a boat. The unfortunate effect of the bow wave is to increase the drag on the body. Experiments with streamlined (torpedo-shaped) bodies have shown that, because of the bow wave, considerably more drag acts on them when they move just below the surface, or even half submerged, than when they travel at the same speed well below the surface.

The height of the bow wave and the extra drag it brings depend on the body's size and speed. A rule that we have already encountered will help us to understand this. In Chapter 2 we learned that, if gravity is important, the motions of different-sized bodies cannot be dynamically similar unless their Froude numbers are equal. Gravity is important in the dynamics of water waves (it tries to flatten them), and so, for bow waves to be dynamically similar, the ships (or ducks,

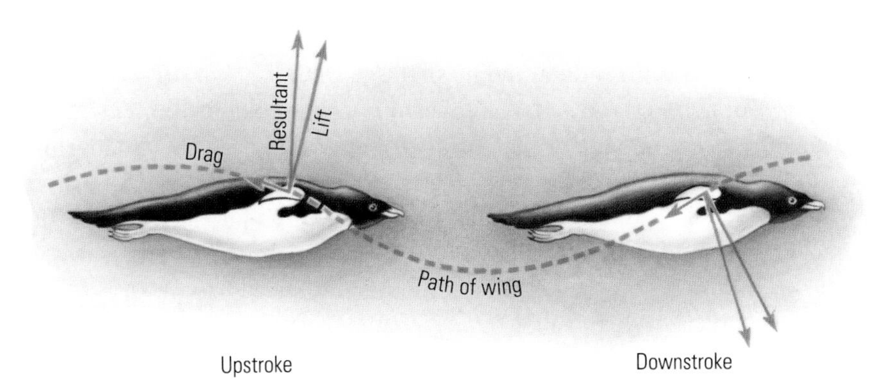


As a penguin swims, the hydrodynamic lift on the wing acts forward and upward on the downstroke, forward and downward on the upstroke.

A Black duck (*Anas rubripes*) swimming, showing the bow wave.

or whatever) responsible for them must travel with equal Froude numbers. In this case,

$$\text{Froude number} = \frac{(\text{speed})^2}{\text{hull length} \times \text{gravity}}$$

The bow wave builds up higher as a ship goes faster. The drag it creates increases sharply at Froude numbers above 0.2 and reaches a maximum at a Froude number of 0.5. Ducks generally swim too slowly for bow wave resistance to be much of a problem. For example, Mallard ducks on a pond had "hulls" about 0.3 meter long and swam at about 0.5 meter per second, so their Froude numbers were only about $(0.5)^2/(0.3 \times 10) = 0.08$. (The gravitational acceleration is about 10 meters per second squared.) Mallards made to swim faster in laboratory tests seemed unable to sustain speeds above 0.7 meter per second (Froude number 0.16). Although ducks presumably could swim faster if they had larger leg muscles, a lot of extra muscle would be needed to give a little more speed, because drag increases so sharply with Froude number. Speedboats reach high Froude numbers by hydroplaning (skimming over the surface of the water), but no animals do that.

The history of boats suggests that the penguin style of swimming may be better than the duck style. Paddlesteamers propelled by drag, like rowboats and ducks, were once common, but they have long since been replaced by screw-driven ships propelled by lift, like penguins. This made it seem reasonable to suppose that the paddling of ducks would use more power than penguin swimming by hydrofoil action at the same speed. To find out whether it does, Russell Baudinette and Peter Gill of Flinders University, Australia, measured the rates of oxygen consumption of swimming ducks and penguins. They trained the birds to swim in a flume, a channel through which water could be driven in a smooth, even stream. The bird was enclosed in a plexiglass chamber, which was suspended just above the surface of the water, so it had to breathe the air in this chamber. Air drawn through the chamber was analyzed to find out how much oxygen the bird was using, and the rate of use of energy was calculated from the oxygen consumption.

Baudinette and Gill compared birds of very similar masses, 1.1-kilogram Black ducks *(Anas superciliosa)* and 1.2-kilogram Little penguins *(Eudyptula minor)*. (This is the Australian Black duck, a different species from the American bird of the same name.) When resting

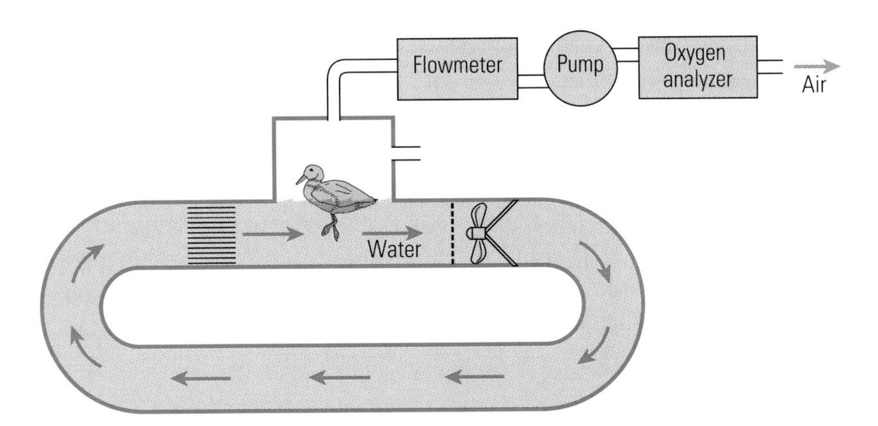


The oxygen consumption of swimming birds can be measured by analyzing the loss of oxygen in a small plexiglass chamber fitted over the bird.

on the surface of still water, both metabolized at rates of about 6 watts per kilogram body mass. Both species could maintain speeds of up to 0.72 meter per second, and they could hold those speeds long enough for it to seem clear that their swimming was powered by aerobic metabolism. When both species swam on the surface at that speed, the ducks used about 18 watts per kilogram and the penguins only 13 watts per kilogram. To calculate the power actually used for swimming, we must subtract the 6 watts per kilogram used while resting on still water: thus the energy cost of swimming is 12 watts per kilogram for the ducks and 7 watts per kilogram for the penguins, at the same speed. The difference may stem in part from the better streamlining of the penguins, but it also seems clear that the penguins' wings, beating up and down through a wide angle, must have pushed on more water than the ducks' feet. We already know that the same thrust can be obtained at less energy cost by pushing a lot of water slowly than by pushing a little water fast.

The wing movements of the penguins were fast compared to the paddling movements of the ducks. Whereas the penguins beat their wings at 3 to 5 cycles per second, the ducks paddled at only 1.5 to 3 cycles per second. It may be interesting to see if these rates of movement are similar to those of equivalent birds flying or walking. When Black ducks walk, they take about 2 strides per second, and when Mallard ducks of similar mass fly (I do not have the data for Black ducks), they beat their wings at about 5 cycles per second. Thus swimming penguins beat their wings at about the same frequency as flying birds of equal mass, and swimming ducks paddle at about the frequency of their walking strides.

The ducks always swam on the surface, but in some experiments the penguins were persuaded to swim well below the surface, visiting it only to breathe. They then produced no bow wave, and the energy cost of swimming was less. At 0.72 meter per second (the maximum speed for sustained surface swimming), they used only about 9 watts per kilogram, considerably less than the 13 watts per kilogram required for surface swimming. After subtracting the 6 watts per kilogram used when the penguins were resting on still water, we find that the energy cost of their swimming movements must have been about 3 watts per kilogram under water and 7 watts per kilogram on the surface. It is much more economical for penguins to swim well below the surface. They can also swim a little faster under water: they could sustain speeds up to 0.84 meter per second, instead of 0.72. For ducks, however, surface swimming may be the more economical option because the air trapped under their feathers makes them so buoyant that they have to paddle hard to keep themselves submerged.

## The Hydrofoil Tail

Dolphins also swim by up-and-down movements of hydrofoils, but their hydrofoils are tail flukes, not wings. The tail beats up and down as the dolphin swims, taking a wavy path through the water. Its angle of attack is presumably adjusted to give upward, forward lift in the downstroke and downward, forward lift in the upstroke, just like penguin wings.

Dolphins often swim close to the bows of ships, staying with a ship for hours even when it is traveling quite fast, at up to at least 10 meters per second (20 knots). This won them a great reputation for speed and stamina, but it has since been realized that they get help from the ship and could not by themselves keep swimming for long at the same speed. Their trick is to swim in the bow wave that is being pushed along by the ship.

Dolphins are propelled by the lift on their tail flukes, much as penguins are propelled by lift on their wings.

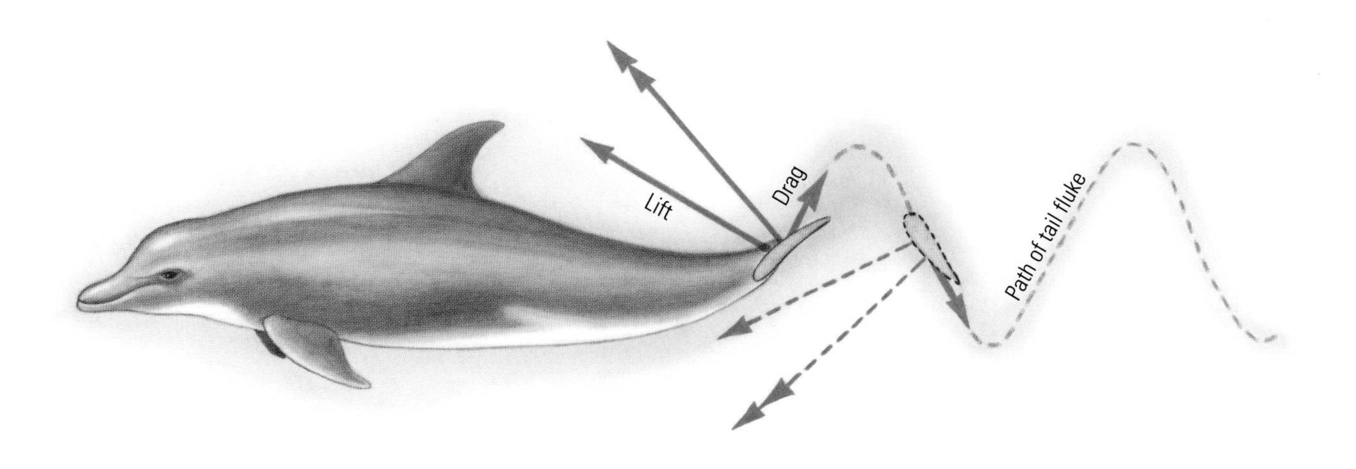

The speeds at which dolphins and porpoises can swim unaided have been measured in experiments with trained animals. The highest speed was recorded by a spotted porpoise (*Stenella attenuata*) that had been trained to chase a yellow wooden lure that was towed by a winch across a lagoon in Hawaii. The porpoise reached 11 meters per second, the speed of an Olympic-class sprinter at the fastest stage of a 100-meter race. However, it could maintain that speed for only a few seconds.

A lot of power is needed to propel so large an animal, at such a high speed. The porpoise was 1.9 meters long, with a mass of 53 kilograms. It was calculated that to propel a rigid body of the same size and shape at the same speed, 2900 watts would be needed. This is a minimum estimate of the power required, because the calculation ignored the kinetic energy given to water in the wake and the likely effects on drag of the beating of the tail. I will show later in this chapter that the energy needed to propel a fish is more than would be needed for a rigid body of the same size and shape, traveling at the same speed.

We will compare that estimate of porpoise power output with the maximum power output of a human athlete. The power produced by athletes has been measured by means of a bicycle ergometer (an exercise bicycle with instruments for measuring work and speed) fitted with hand cranks so that the athlete's arm muscles could do work, as well as the leg muscles. The greatest power that a 70-kilogram athlete could maintain for a few seconds was only 1400 watts, about half the estimated power output of the 53-kilogram porpoise.

Porpoising is a well-known habit of dolphins and porpoises. They swim along, alternately swimming below the surface and leaping clear of it. This looks strenuous and playful, but it has been suggested that porpoising may reduce the energy cost of fast swimming. The idea is that because air is so much less dense than water, the drag on a body moving through air is much less than the drag on a body moving at the same speed through water. Leaping costs energy, of course, but the suggestion is that if the animal is traveling fast enough, the energy cost of the leap is outweighed by the energy saved by escaping from the drag of the water.

Scientists argue about the energy cost of the leap. One opinion is that the principal energy cost is the kinetic energy associated with the

upward component of the animal's velocity as it leaves the water. Against this view, it has been pointed out that although the dolphin is traveling downward when it reenters the water, its kinetic energy then is almost the same as when it left the water traveling upward, so no energy is needed to give kinetic energy to the animal itself. However, like any body moving through water, a dolphin pulls along with itself a thin coating of water, called a boundary layer. When the animal leaves the water, this boundary layer falls off (photographs show water falling away from leaping dolphins) and its kinetic energy is lost. Whether the work of the leap is the upward kinetic energy of the dolphin itself or the kinetic energy of the lost boundary layer or both, it is proportional to the square of the speed. From the theory of projectiles, we know that the length of the leap also is proportional to the square of the speed: if dolphins traveling at different speeds leave the water at the same angle, they will leap four times the distance for twice the speed. Since both the length of the leap and its likely energy cost are proportional to the square of speed, the energy cost *per unit distance* of leaps is the same for all speeds. The higher the speed,

Bottlenose dolphins (*Tursiops truncatus*) porpoising. The water of the boundary layer disperses as spray when they leap.

A school of tuna, photographed in the Sea of Cortez. Their streamlined bodies and tall narrow tail fins are beautifully adapted for swimming.

however, the higher the drag in water; thus the energy cost per unit distance swum in water increases with increasing speed. This argument seems to tell us that there must be a speed above which leaping saves energy. The doubts about the energy cost of leaping make it difficult to estimate what that speed may be, and few measurements have been made of the speeds at which dolphins and porpoises do or do not leap. Therefore it is difficult to test the hypothesis that porpoising saves energy in fast swimming.

Penguins also porpoise, and their porpoising habits have been studied more systematically by Clifford Hui of the University of California, Los Angeles. He found that penguins porpoised at speeds of 3 to 4.5 meters per second, but seldom at lower speeds, which seems to fit the energy-saving hypothesis. The leaps were short, however, never measuring more than 22 percent of the total distance traveled, and so any energy savings must have been small. Hui observed that the penguins were out of the water for about 0.3 second in each leap, just a little longer than a walking penguin needed for each intake of breath. He suggested that porpoising saves energy by giving the bird opportunities to breathe without having to swim at the surface where drag would be increased by the bow wave.

Tunnies and similarly shaped fishes use their tail fins as hydrofoils in essentially the same way as dolphins use their flukes. The principal difference is that dolphins have horizontal flukes, which beat up and down in swimming, whereas tunnies have vertical tail fins that beat from side to side: the swimming action of a tunny is like that of a dolphin lying on its side.

Tunnies and their relatives have a reputation for speed, and attempts have been made to find out how fast they can swim by letting them run while attached to an instrumented fishing line. A fish was hooked, then allowed to pull the line out freely, and the instruments recorded the rate at which the line unwound. This method gave an astonishing speed of 21 meters per second for a wahoo (*Acanthocybium solanderi*) and for another species of tuna. This speed is so much greater than the highest speeds recorded by any other method, for any swimming animal, that I suspect an error. As the line was drawn out, the instruments sensed magnetic markers spaced at intervals along it, and each sensing produced a blip on a strip of paper in a recording instrument. From the number of blips per unit time, it was a simple matter to calculate speed. It seems possible that the record was corrupted by electrical "noise" and that some of the blips that were counted as showing the passage of markers were merely noise. The highest speed recorded for a fish by any other means seems to be 10.5 meters per second for a 0.5-meter skipjack tuna (*Katsuwonus pelamis*).

The lift on a tunny's tail (or a dolphin's fluke or a penguin's wing) propels the animal, but the drag on the tail (or fluke or wing) slows it down. A tail that supplies lift needs a special shape to minimize drag. We saw in Chapter 4 that for the highest ratio of lift to drag, wings should have high aspect ratios—they should be long and narrow. Similarly, we expect tunny tail fins to be tall and narrow, and in fact they have much higher aspect ratios than the tail fins of most other fish.

## Jet Propulsion

We will return soon to the swimming of fishes, but first we will look at jet propulsion, a swimming technique used by squids and a few other animals. To drive themselves along, these animals squirt water out of cavities in their bodies. A jetting squid does not move steadily like a jet aircraft but in a series of jerks, speeding up as it squirts water out and slowing down as it takes in the water that will be squirted out in the next jet. Medium-sized squids of about 0.4-kg body mass make about one jet per second when they swim fast.

The mechanism for squirting water is a variant of the mechanism squids use to pass water through the gills. The gills of squid are hidden inside the body in a cavity enclosed by a muscular wall. The wall is called the mantle, and the cavity it encloses is called the mantle cavity. Fish breathe by drawing water in through the mouth and blowing it out through the gill covers; on the way it passes over the gills. Similarly, squid draw water into the mantle cavity through one opening and blow it out through another, so that it passes over the gills on the way. The water enters through a slit across the width of the belly and leaves by way of a round opening, the funnel. (Valves prevent flow in the reverse direction.) Gentle rhythmic contractions of the mantle muscles are enough to keep water flowing over the gills for respiration, but more forceful contractions eject a rapid jet through the funnel and can propel the animal at high speed. If the funnel is in its resting position, pointing toward the animal's head, the jet propels the animal backward, but the funnel can be curled around to point backward and drive the animal forward.

I do not know how fast squid can swim. One that was 0.2 meter (8 inches) long was filmed accelerating backward from rest to 2.1 meters per second by a single contraction of its mantle. It presumably could have reached higher speeds by making several contractions in rapid sequence. Oceanic "flying squids" swim fast enough to leap quite high out of the water, and sometimes they even land on the decks of ships. If the heights or lengths of these leaps were known, that information could be used to calculate the speeds, but no one seems to have taken the right measurements.

We used measurements of oxygen consumption to compare the energy costs of the paddling of ducks and the hydrofoil swimming of penguins. Now we will use the same method to compare the jet pro-

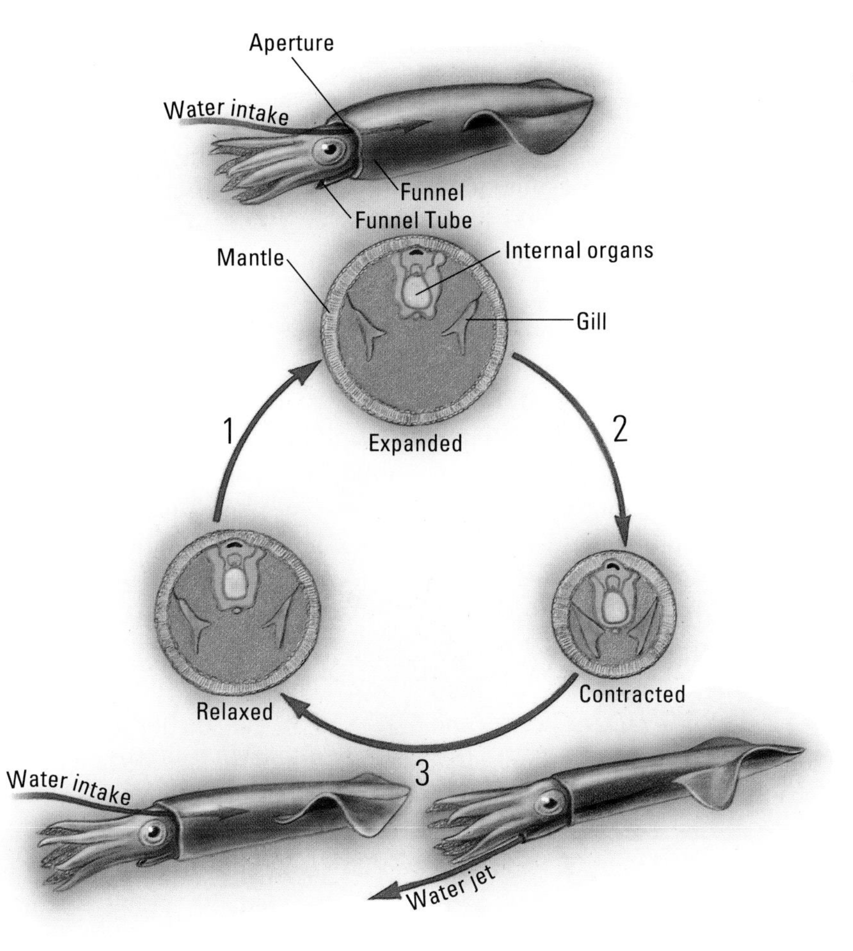


Squid propel themselves by squirting jets of water out of the mantle cavity, which contains the gills. Water is drawn in through a wide slit in the belly and squirted out through the round funnel.

pulsion of squids with the tail-beat swimming of fishes. Dale Webber and Ron O'Dor at Dalhousie University, Nova Scotia, measured the rates of oxygen consumption of swimming squids. They let squids propel themselves in a water tunnel consisting of a loop of large pipes through which water was kept circulating by a powerful pump. The tunnel was designed like a wind tunnel so that the flow was as smooth as possible, and as even as possible across the width of the pipe, in the section where the squid swam. Provided the speed of the current was not too great, trained squid swam steadily against the current for several hours. As the same water circulated repeatedly around the loop, the concentration of dissolved oxygen was gradually reduced (but only a little) by the squids' respiration. An oxygen electrode measured this concentration, and once Webber and O'Dor knew by how much the oxygen was reduced, they could calculate the rates of oxygen consumption at different swimming speeds. The highest speed that could be sustained by 400-gram squid was 0.8 meter per second, but they could swim much faster in a short burst of activity.

Rates of oxygen consumption of swimming fishes had been measured previously by the same method, which was pioneered by Roland Brett of the Fisheries Research Board of Canada. Webber and O'Dor compared the oxygen consumption of squid with that of young Pacific salmon. For the comparison to be valid, they had to ensure that the animals to be compared were of similar size. Their squid had a mass of 400 grams when the mantle cavity was empty, but while swimming had an average of about 100 grams water in the mantle cavity: for that reason, the squid were compared with 500-gram salmon. At its maximum sustained speed of 0.8 meter per second, the squid's metabolic rate was 1.6 watts greater than at rest. The salmon could sustain a speed of 1.4 meters per second and even at that higher speed used only 1.2 watts more than when resting. Thus the squid comes very badly out of the comparison, which may seem odd since it looks as beautifully streamlined as the salmon.

Part of the reason seems to be that squid can squirt out only as much water as the mantle cavity will hold, about 200 grams of water for a 400-gram squid. (The 100 grams mentioned above was the *average* content of the cavity over a cycle of expansion and contraction.) In contrast, fish can push with their tails on very large masses of water. I have repeatedly made the point that less energy is needed to supply the same thrust when large masses are accelerated to low speeds than when smaller masses are accelerated to higher speeds.

In an ingenious experiment, Charles McCutchen of the National Institutes of Health, Bethesda, Maryland, measured the masses of water on which fish push. He half-filled an aquarium with cold water, then topped it up very carefully with warmer water so that the warm floated on top of the cold, mixing with it very little. When a fish swam at the interface between the warm and cold water, it stirred some of the cold up into the warm and some of the warm down into the cold. Because the refractive index of water (the extent to which it bends light rays) depends on the temperature, McCutchen was able to make the disturbance in the water visible by suitable lighting. Thus he was able to film the movements of the water, as well as those of the fish.

A typical film sequence from his experiments shows a small fish coasting along, then making two strong tail beats that accelerate it and send it off in a new direction. The tail beats set two masses of water moving, one to the left and one to the right. It is impossible to measure the masses of moving water directly from the film, but here is how McCutchen calculated them. He knew the mass of the fish. The film gave the initial and final velocities of the fish and the velocities of the masses of water moved by the two tail beats. (Velocity means speed *and direction* of movement.) A fundamental physical principle, the Principle of Conservation of Momentum, told him that the change in the momentum (mass multiplied by velocity) of the fish must be equalled by the change in the momentum of the water. By

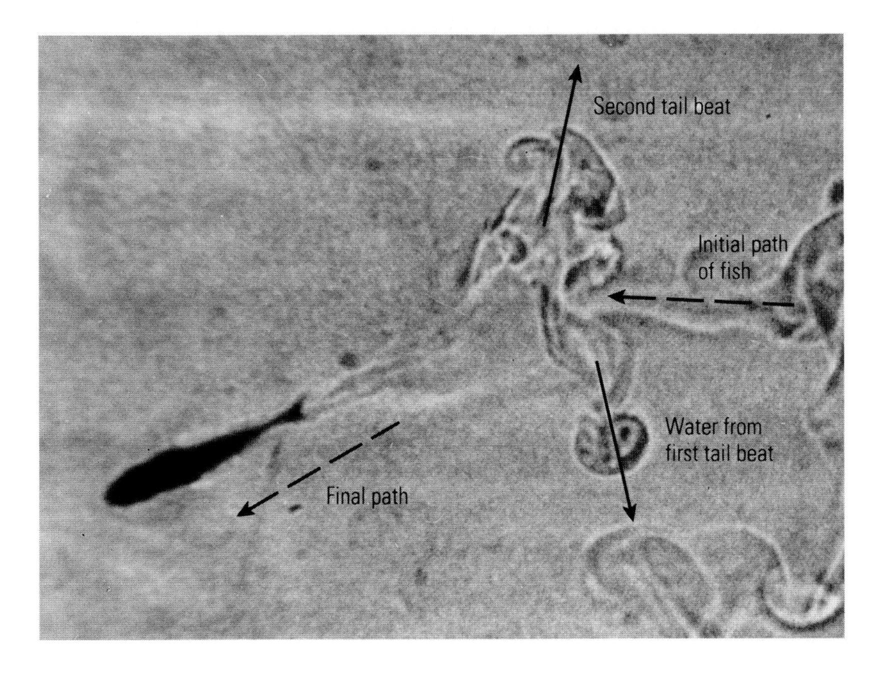


The water movements in the wake of a small fish, made visible in an experiment by Charles McCutchen.

applying this principle, he was able to calculate that the two masses of water were each about three times the mass of the fish. Each tail beat of the fish pushes on water amounting to three times its body mass. In comparison, each contraction of the squid's mantle squirts out water amounting to only about half the body mass. This difference makes the swimming of fish much more efficient than the jet propulsion of squids.

## Undulating Swimmers

We have discussed rowing, hydrofoil swimming, and jet propulsion. The fourth and last major swimming technique is undulation, the technique used by eels. The movements are like the serpentine crawling of snakes: the body is thrown into waves that are made to travel backward along it. In serpentine crawling the waves remain stationary *relative to the ground* as the body moves forward, but in the eels' style of swimming the waves travel backward through the water as the body moves forward.

The mechanism was first explained clearly in 1933 by James Gray of the University of Cambridge, England, one of the pioneers of studies of animal locomotion. He took a film of an eel swimming and superimposed outlines traced from successive frames. Then he selected one short section of the body and highlighted its position in successive outlines. His analysis showed that, whether the segment was moving toward the left or toward the right, it was always angled to push water backward.

An anemonefish (*Amphiprion*) among the tentacles of a sea anemone. Its slow, precisely controlled swimming is powered by undulations of the fins.

Not only eels but also some worms swim by undulating their bodies. Cuttlefish swim fast by jet propulsion and most true fish swim fast by tail movements, but to make slow swimming movements cuttlefish and many fish undulate their fins. A fish swimming in this way keeps its body straight and its main swimming muscles inactive: only the fins undulate, powered by small muscles at their bases.

This style of swimming makes some teleost fishes wonderfully maneuverable. In addition to tail fins, they have fins on their backs, on their bellies, and on their sides. They can send waves backward along their fins to drive themselves forward, or they can send waves forward to drive themselves backward. They can send waves upward or downward in the tail fin and the fins on their sides to make them-

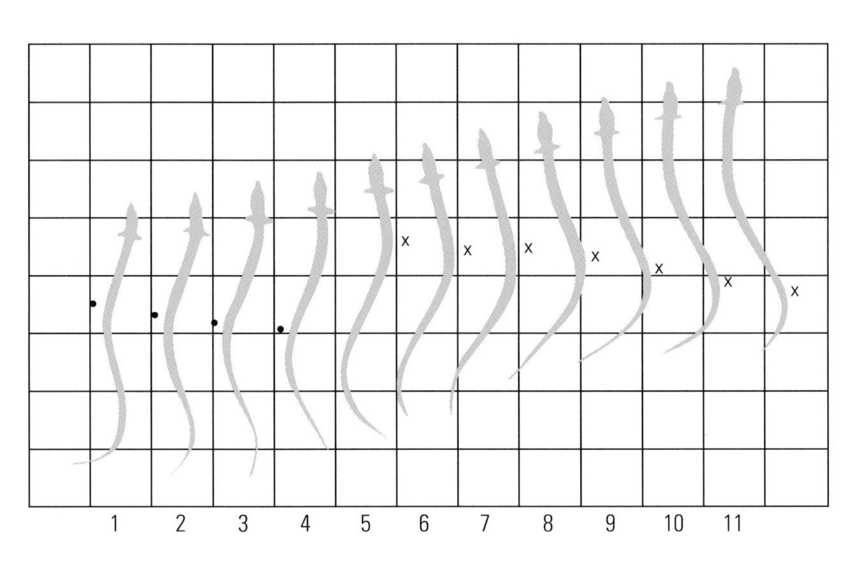


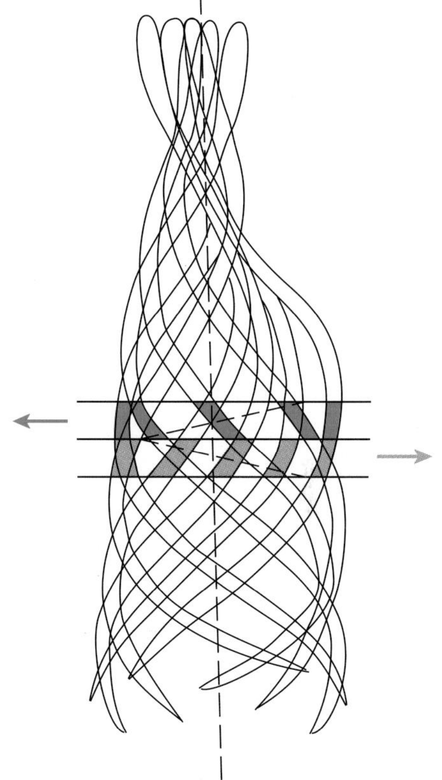

An eel *(Anguilla anguilla)* swims by undulating its body. In these illustrations from James Gray's pioneering study in the *Journal of Experimental Biology,* outlines traced from successive frames of a film show the sequence of movement *(left),* and the same segment is highlighted in the nine superimposed outlines *(right).* The upper highlighted segments are moving to the left and the lower segments to the right.

selves move vertically up or down. A fish swimming by beating its tail is like an airplane; it can only travel forward. But a fish swimming by undulating its fins can move forward, backward, up, or down, like a helicopter. This maneuverability is especially useful to fishes that live on coral reefs and feed on coral polyps or on small animals that live in crevices. They have to move very precisely among the clutter of the reef to reach their food.

## The Energy Cost of Swimming

Eels and tunnies represent two contrasting styles of movement. Eels have only small tail fins and swim by undulating their entire bodies. In contrast, tunnies have tall tails that act as hydrofoils; they beat their tails from side to side by bending the rear part of the body, keeping the rest almost straight. The swimming of most other fishes

Teleost fishes can swim forward, vertically, or backward by undulating appropriate fins.

is a combination of the undulatory swimming of eels and the hydrofoil swimming of tunnies. These other fishes have large, hydrofoil tail fins like tunnies, but they also undulate the whole body, like eels.

A fish starting from rest drives water backward: to give itself forward momentum, it must give the water backward momentum. When it swims at constant speed, however, its momentum is constant and so the total momentum of the water must also remain unchanged. Alternate strokes of the tail drive water to left and right, giving it momentum alternately to either side, but the leftward momentum given by one stroke is balanced by the rightward momentum given by the next. If we count leftward momentum as positive and rightward as negative, for example, it is easy to see that the total momentum remains unchanged.

That view of fish swimming is the basis of an influential theory that shows how to calculate the power output of a swimming fish. Its author, James Lighthill, is a remarkable hydrodynamicist who has been at different times director of the Royal Aeronautical Establishment, Farnborough, England; a professor of mathematics at Cambridge University; and provost (that means president) of University College, London. Lighthill showed how the mass and velocity of the water pushed by each tail beat could be calculated if the movements of the fish were known. The force needed to give sideways momentum to the water could also be calculated, and the power output of the muscles moving the tail was this force multiplied by the tail's sideways velocity. A great deal could be calculated if the fish's movements were known.

There is another, more obvious way to calculate the power needed for swimming. A typical fish is a well-streamlined body like a torpedo,

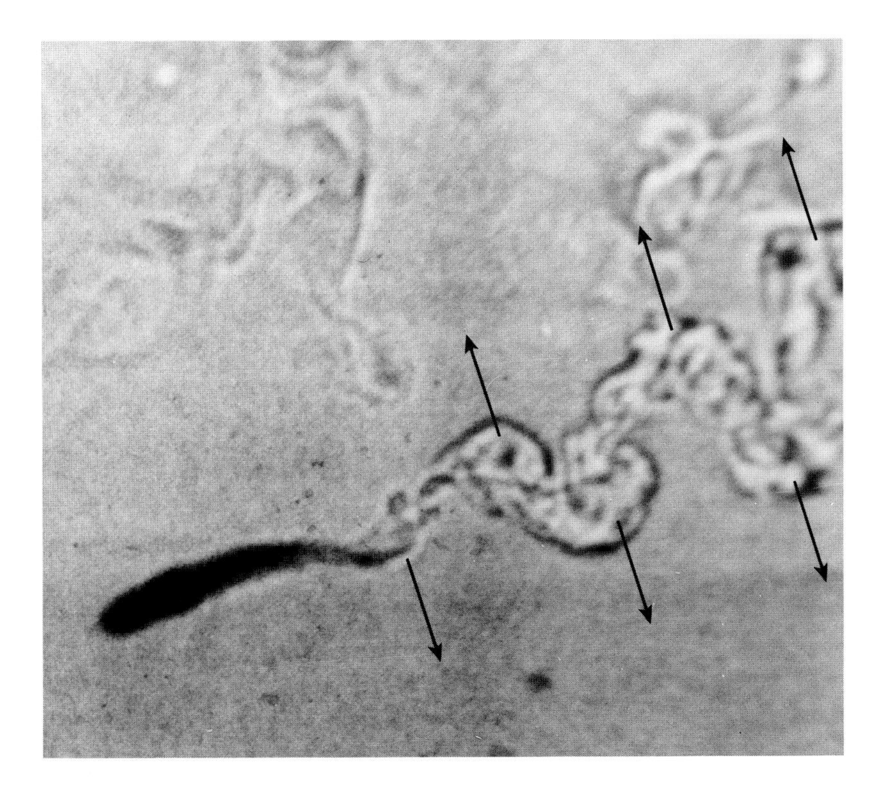

The tail movements of a fish swimming at constant speed drive water to either side, not backward. The movements were made visible in this photograph by Charles McCutchen's method, using warm water floating on cooler water.

and it is easy to calculate the power needed to propel a torpedo of any size at any speed, as hydrodynamics books explain. In a book published in 1967 I had used this simple-minded method to calculate the power needed to propel various fish, whose rates of oxygen consumption had been measured. I compared the calculated power requirement to the actual power consumption indicated by the rate of oxygen consumption. For example, an 87-gram goldfish had used oxygen at a rate corresponding to an energy consumption of about 10 milliwatts when resting and 60 milliwatts when swimming in a water tunnel at 0.64 meter per second. (A milliwatt is a thousandth part of a watt.) The difference of 50 milliwatts was presumably needed for swimming. As explained in Chapter 1, muscles commonly work with efficiencies of about 0.2: about 5 joules of metabolic energy are needed to do 1 joule of work. Thus the 50-milliwatt metabolic energy cost of swimming suggests a mechanical power output of about 10 milliwatts. However, the calculation that treated the fish as a rigid torpedo gave a much smaller estimated power requirement, only 2.4 milliwatts. The fish was using four times more power than seemed to be needed. Are fish muscles unusually inefficient, or is there some other explanation?

About this time, Paul Webb—then a graduate student at the University of Bristol, England, and now a professor at the University of Michigan—was experimenting with trout in a water tunnel, measuring their oxygen consumption and also filming them so that their power output could be calculated according to Lighthill's theory. Webb was also interested in the efficiency of the swimming muscles and devised an ingenious method to measure it. He attached baffles to the backs of some of his fish so that when they swam they had to do work against the drag on the baffle (which could be calculated easily) in addition to the work needed to propel their own bodies. He measured the oxygen consumption of fish swimming at the same speed, with and without baffles. The difference told him how much metabolic power was needed to do the work against the drag on the baffle, making it possible to calculate the efficiency of the muscles. His results give efficiencies of about 0.14, just a little lower than might have been expected.

Webb now had two methods for calculating how much mechanical power was required for swimming without baffles. He could use Lighthill's theory, or he could multiply the metabolic energy cost by the efficiency. The two estimates agreed well, but both were about three times higher than the result of the torpedo calculation. Webb concluded that the Lighthill theory gives a much better estimate of power requirements than the simpler and more obvious one.

A trout with a baffle attached, for an experiment by Paul Webb.

There seem to be two reasons why the torpedo calculation underestimates the energy cost of swimming. First, it ignores the kinetic energy given to the water, which is pushed to either side as the tail beats. This is not very important, however, because the large masses of water that are pushed move with only low sideways velocities. Second, and much more important, the calculation treats the fish as a rigid torpedo, but it is not rigid. The undulating movements of swimming fish probably thin the boundary layer, increasing drag on the fish's body.

## The Swimming Muscles of Fish

Like other vertebrate animals, fish have red aerobic muscle fibers and white anaerobic ones. In legged vertebrates, some muscles are predominantly aerobic and others anaerobic, but the two types of fiber are often mixed in the same muscle. In fish, however, they are sharply separate. A photograph on page 15 shows that the red aerobic fibers form distinct bands along either side of the fish. The separation of the two types of muscle makes it relatively easy to find out when the fish uses each type. With this information, we will be able to examine an important hypothesis about the functioning of muscles in general, one that concerns the relationship between rate of shortening and power output.

A simple procedure records when a fish uses each type of muscle. In several sets of experiments, very fine wires were implanted in the red and white muscles of fish. (This is not a painful procedure, and it has often been performed without ill effects in experiments on human volunteers.) The fish then were allowed to swim against the current in a water tunnel, with the wires connected to equipment that recorded the electrical impulses in the muscles. Such impulses occur in muscles whenever they are active, so the records showed which muscles were used at different speeds.

These experiments showed that the red muscles alone are used when fish swim slowly and that the white muscles are used only for rapid acceleration or in bursts of speed. For example, in experiments by Larry Rome of the University of Pennsylvania, small carp about 13 centimeters in length used their red muscles alone at swimming speeds below 0.4 meter per second, but brought their white muscles into use at higher speeds. Although the fish could continue swimming indefinitely at speeds below 0.4 meter per second, they could not sustain higher speeds for long. It seems that at speeds below 0.4 meter per second the fish were powered entirely by red muscles working aerobically, so no oxygen debt accumulated, whereas at higher speeds white muscles had to be used and they built up an oxygen debt. Lactic acid accumulates in the muscles and the fish must stop to rest before its concentration becomes intolerable. You may think it would be better for the fish to have more red muscle and less white, so that higher swimming speeds could be sustained, but more red muscle would do no good unless the gills and blood system were also enlarged, the gills to take up oxygen faster and the blood system to transport it to the muscles.

We saw in Chapter 1 that the faster a muscle shortens, the less force it can exert. Its power output (the rate at which it does work) is the force multiplied by the rate of shortening. The power is obviously zero when the muscle is not shortening, but it is also zero when the muscle is shortening at the maximum rate, because the force is then zero. Power output is at its maximum at an intermediate speed, about one third of the maximum shortening speed. The efficiency of muscle (work output divided by metabolic energy input) is generally also maximal at about this rate of shortening. Different muscles have different maximum rates of shortening, and we may expect to find that these maximum rates are adjusted so that the rates of shortening in normal use, in movements in which maximum power is needed from the muscles, are around one third of the maximum.

No one had checked whether any muscle's normal rate of shortening was in fact one third of the maximum until Larry Rome decided to try to do it for the swimming muscles of carp (*Cyprinus carpio*). He recruited a team (myself included) and we assembled for the experiments at the Wood's Hole Marine Biological Laboratories, Cape Cod. Since carp live in fresh water there was no particular advantage in being near the sea, but we chose Wood's Hole because Marvin Freadman had a suitable water tunnel there.

I have already described the experiment that showed that the carp swam by using only their red muscles, at speeds of up to 0.4 meter per second. We formed the hypothesis that at that speed, when the power output of the red muscles had to be highest, they would be found to be shortening at one third of their maximum rate.

To obtain the information needed to test the hypothesis, we took high-speed films of the carp swimming in the water tunnel at this and other speeds. We filmed from above, looking down on the fish so that the film showed how much its body was bent at each stage of the tailbeat cycle. We analyzed the film, measuring the curvature of the backbone at each stage.

We still had to find out how much shortening of the red muscles was needed for each degree of curvature of the backbone. We were able to work this out by geometry, but we checked the calculation by means of a simple experiment. The sarcomeres from which muscle fibers are built (see the diagram on page 8) can be seen under a microscope as a pattern of stripes across the fibers. The more a fiber shortens, the narrower are the stripes. We killed a few carp and fastened them to wooden boards with their bodies bent to various degrees. We left them until rigor mortis had set in, tightening all the muscle fibers, then treated them with a chemical fixative so that fiber lengths would not change further. We dissected out some of the red fibers, examined them under a microscope, and measured the spacing of the stripes—that is, the length of the sarcomeres. This procedure confirmed that the geometric analysis had been accurate.

We were now able to calculate how much the red muscle fibers must have shortened to have bent the fish to the curvatures shown in our films. We found that in swimming at 0.4 meter per second (the maximum rate that could be sustained by the red muscle alone), these muscles had to shorten at a rate of 1.7 muscle lengths per second. In experiments with bundles of red muscle fibers, Rome had shown that their maximum rate of shortening was 4.7 lengths per second. As we had predicted, when the muscles needed to produce maximum power, they were shortening at about one third of their maximum shortening speed. The muscles were shortening at the rate at which they could produce most power and at which they probably worked most efficiently.

The red fibers of fish have a simple arrangement: they run lengthwise along the sides of the body. The white fibers, however,

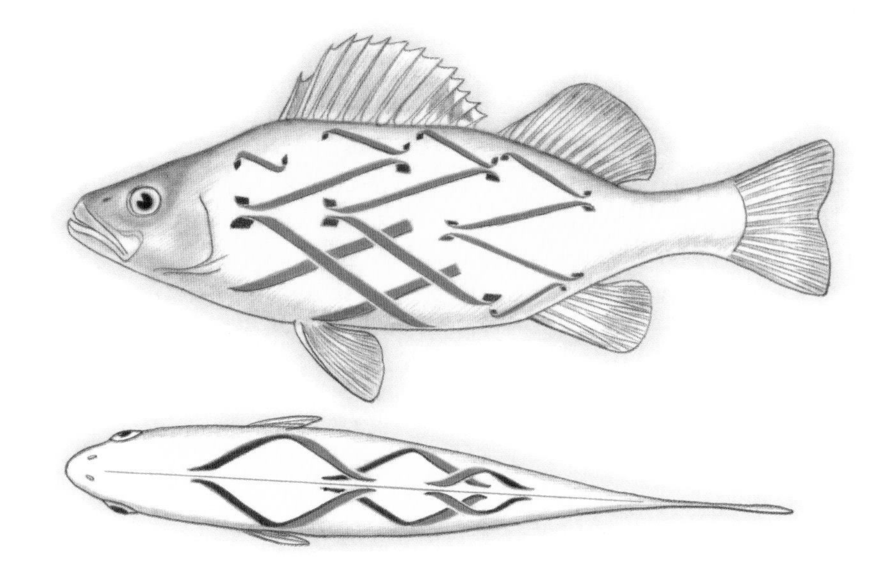

The helical arrangement of the white muscle fibers in a typical teleost fish.

are arranged in more complicated ways that have important consequences. Different patterns are found in different fish, but the commonest pattern in teleost fishes, and the one found in carp, has white fibers running at angles of up to 35 degrees to the long axis of the body. The muscle is partitioned into segments (which separate easily after cooking) and each fiber runs only the length of a segment, from one partition to the next. However, if you follow a series of fibers, connected end-to-end through the partitions, you will find a pattern: these chains of fibers run helically, like the strands of a rope. Their helical structure has important consequences for muscle contraction.

Imagine that the fibers were not so arranged, but instead all ran parallel to the long axis of the body. Imagine the fish bending to such an extent that the outermost fibers, just under the skin of the sides of the body, had to shorten by 10 percent. Fibers halfway between the skin and the backbone would have to shorten by only 5 percent, and fibers alongside the backbone would shorten hardly at all. In each tail beat, the outermost muscle fibers would have to shorten quite a lot and quite fast, whereas the innermost fibers would shorten much less in the same time and therefore more slowly.

Now consider how the actual arrangement of white fibers affects the shortening of the muscles. As we have seen, chains of fibers run helically like the strands of a rope. Each chain is close to the backbone for parts of its course and nearer the skin of the fish's side for others. The result is that when the fish bends, say, to the right, all the white fibers on the right side have to shorten by about the same percentage of their length. A bend that required 10 percent shortening of the red fibers would require only about 2.5 percent shortening of the white. A corollary, of course, is that 10 percent shortening of the white fibers would bend the fish into a very much tighter curve than would 10 percent shortening of the red.

Our carp started using their white muscles at speeds around 0.4 meter per second, but they could swim short bursts at very much higher speeds. (The mass of the white muscle that powers fast bursts is many times the mass of the red.) The fish probably could have managed short bursts at up to about 1.6 meters per second. A fish swimming at this speed makes almost the same movements as it does at 0.4 meter per second, and it moves forward about the same distance in each tail beat, but it makes about four times as many tail beats per second. If the same muscles powered swimming at 1.6 meters per second as at 0.4 meter per second, they would have to shorten about four times as fast. However, the helical arrangement of the white fibers means they have only to shorten by one quarter as much as the red ones to cause the same movement. Thus, at a swimming speed of 1.6 meters per second the white fibers should only have to shorten as fast as red fibers shorten at a speed of 0.4 meter per second—1.7 lengths per second. Rome measured the maximum rates of shortening of bundles of white muscle fibers and found they were about 13 lengths per second. Even in swimming at 1.6 meters per second, the white fibers would probably have to shorten at only about one eighth of their maximum rate, well below the rate of shortening at which they would produce maximum power.

At first sight this seems puzzling and suggests a fault in fish design. The most probable explanation is that the highest power outputs are needed not when the fish is swimming fastest, but when it is accelerating suddenly from rest. The body bends to much stronger curves, and the muscles have to shorten much more, in fast starts than in steady swimming. The red fibers probably cannot shorten enough to bend the fish to the curves seen in fast starts, but because of their helical arrangement, the white fibers can. High-speed films of startled carp showed how rapidly they bent, and we calculated that the rate of shortening of the white fibers was about 5 lengths per second, just a little more than one third of their maximum shortening speed. The maximum shortening speeds of the two types of muscle seem to be adjusted so that the red muscle is delivering its maximum power at the highest speed that it can power, and the white muscle delivers its maximum power in fast starts.

This understanding of fish muscle function has been confirmed and extended by the work of John Altringham and Ian Johnston at St. Andrews University, Scotland. They also performed experiments on bundles of muscle fibers, but instead of simply letting the fibers shorten at different speeds they repeatedly stretched them and allowed them to shorten, simulating their action in swimming fish. They used muscles from bullrout *(Myoxocephalus scorpius),* a peculiarly ugly marine fish. They found that red muscle fibers gave maximum power output when shortening and being stretched at a rate of 2 cycles per second, and white fibers at about 6 cycles per second. These rates seem to match the highest tail-beat frequencies that can be powered by the two types of muscle.

The red muscle seems beautifully adapted for swimming at the highest speed it can power and the white muscle for fast starts, but that leaves a problem. If a fish swims just a little above its maximum aerobic speed, it will have to use its white muscle even though that muscle will be shortening far more slowly than the rate at which it is most efficient. At least some species solve this problem by "burst and coast" swimming, alternating short bursts of rapid tail beats with periods of gliding with the body straight. During the rapid tail beats, the white muscles may well be shortening at near-optimal rates, and thus this form of travel may make it possible to use the white muscles efficiently, even at quite low speeds. (Notice how similar this explanation is to the explanation of bounding flight in Chapter 5.) Paul Webb found that 0.2-meter trout and musky (a hybrid of two *Esox* species) swim continuously at speeds up to 0.4 meter per second (presumably using their red muscles), but often use burst-and-coast swimming at speeds of 0.4 to 0.9 meter per second.

This chapter has shown the many ways in which animals swim: by rowing, by means of hydrofoils, by jet propulsion, or by undulation. It has shown how penguins swimming by hydrofoil action of their wings use less energy (at the same speed) than ducks paddling with their feet. It has also shown that the tail-beat swimming of fish is faster and more economical than the jet propulsion of squids. Finally, it has shown the very beautiful way in which evolution has arranged the swimming muscles of fish to obtain the best possible swimming performance.

# Small-Scale Locomotion

Comb jellies (ctenophores), some of them several centimeters long, are the largest animals to swim by beating cilia. Most other creatures that use this means of propulsion are of microscopic size.

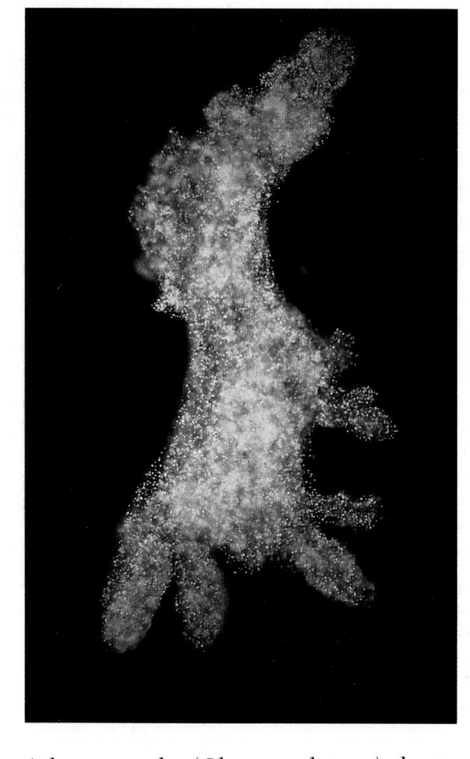

A large amoeba (*Chaos carolinense*) about one millimeter long.

The movements described in previous chapters are all driven by muscles, but very small organisms are powered by other motors. Many of these tiny creatures can be seen only through a microscope, and most have bodies that consist only of a single cell.

The bodies of animals are divided into cells, each controlled by a nucleus whose chromosomes carry a full set of genes, the inherited instructions for the processes of life. Typical cells have diameters between 10 and 100 micrometers (between one hundredth and one tenth of a millimeter).

The smallest living things consist of just one cell. We are not concerned in this chapter with the smallest living things of all, but mainly with the protistans, most of which are similar in size to the single cells of many-celled animals. Some protistans (for example, amoebas) are traditionally thought of as animals, and others (such as diatoms) as plants, but the current fashion is to put all the protistans into a kingdom Protista, separate from both the animal and the plant kingdoms.

A few of the protistans are much larger than most cells. The giant amoeba *Pelomyxa* is about 2 millimeters long, much too large to be controlled easily by a single central nucleus: it has many nuclei, scattered through its single-celled body. Some ciliate protozoans such as *Spirostomum,* belonging to the group that we call the ciliates, are as much as 1 millimeter long, but in addition to an ordinary nucleus they have a long macronucleus extending through most of the length of the body and containing multiple copies of the genes.

This chapter describes the locomotion of these single-celled organisms, as well as that of some small many-celled animals that move

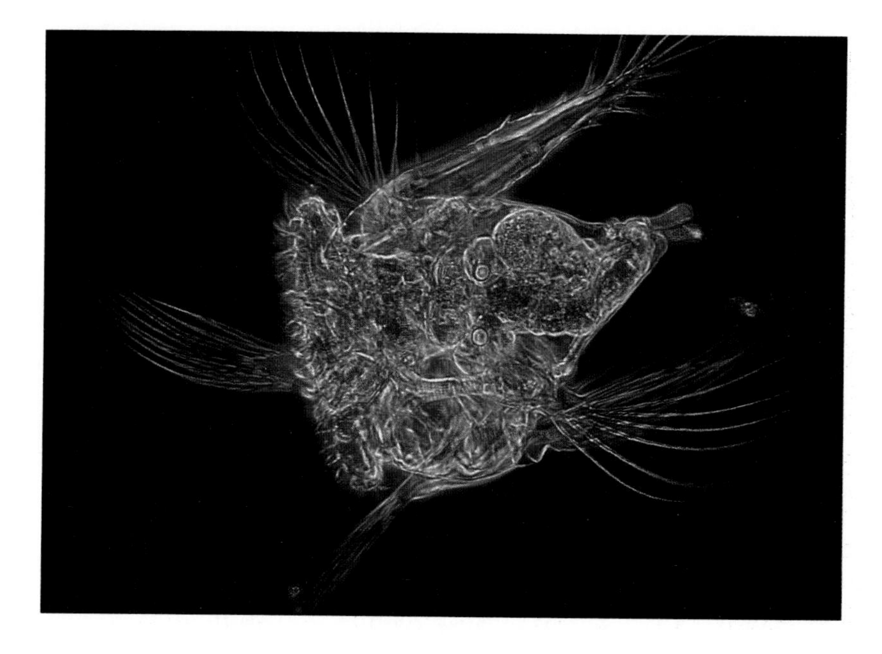

A living rotifer *(Hexarthra),* about 0.2 millimeter long. The cilia can be seen as a short fringe around the flat end of the body (on the left). The much longer bristles are not cilia, and cannot beat like them.

in similar fashion. They include rotifers, which are common in ponds. Rotifers are similar in size to ciliate protozoans but have bodies made up of large numbers of tiny cells: about 1000 in the case of *Epiphanes,* which is about 0.5 millimeter long.

## The Crawling of Amoebas

The amoebas are the most famous of the protistans. They look very simple under the microscope: the observer sees a transparent, irregularly shaped blob, and within that blob are visible several structures—a nucleus (or several), a few vacuoles containing food that is being digested, and a pulsating structure called the contractile vacuole, which pumps excess water out of the body. Although there is no permanent body part designed for locomotion, amoebas can crawl, but only slowly. The highest speed of an amoeba is only 5 micrometers per second, or 18 millimeters per hour.

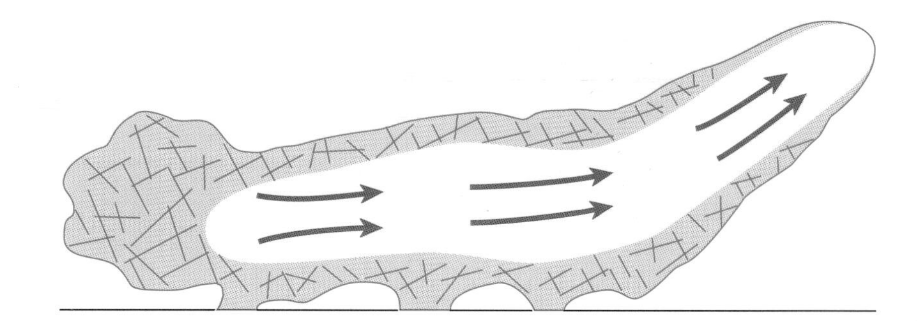

As an amoeba crawls (seen in profile), jellylike ectoplasm becomes fluid endoplasm at the rear end, and endoplasm becomes ectoplasm at the front.

This slow movement happens in a way that looks mysterious. A lobe (called a pseudopod) extends from the side of the body in the direction of motion, and the body contents flow into it. This empties the rear end of the body, which retracts so that the whole cell moves forward. We can see the flow of body contents by following the movements of the various kinds of particles that float within the amoeba. The core of the body (the endoplasm) flows like a liquid, with its particles swirling around freely. The outer layer (the ectoplasm), however, seems to be jellylike and anchored to the ground: particles in it remain stationary relative to each other and to the ground. New ectoplasm must form at the front of the amoeba as the pseudopod extends, and ectoplasm must be lost at the rear. Presumably the old ectoplasm at the rear is transformed into endoplasm, which flows forward and forms new ectoplasm at the front.

Although we cannot explain the movement of amoebas by pointing to structures that we would call muscles, these organisms do contain the major components of muscle. In Chapter 1 I explained how muscle contraction depends on the interaction between thick filaments of the protein myosin and thin filaments that consist largely of the protein actin. Amoebas contain similar proteins—so similar that myosin from vertebrate muscles attaches readily to amoeba actin.

The thick filaments of muscle are bundles of myosin molecules, each molecule shaped like a letter Y with a very long stem. The stems of the Y's form the filament itself and the arms project to form the crossbridges. The myosin of amoebas is not bundled into filaments, and the molecules are not Y-shaped. Each molecule consists of a single arm, similar to a muscle crossbridge, and a short stem.

Actin filaments, both of muscle and of amoebas, are built from large numbers of individual actin molecules. Amoebas prepared by some methods of electron microscopy contain a dense network of actin filaments, but when they are prepared by other methods no filaments can be seen. This seems rather puzzling but suggests that the filaments may be capable of breaking down and re-forming. It seems likely that the ectoplasm of living amoebas is stiffened by a network of actin and that the actin network breaks down into individual molecules when the ectoplasm becomes fluid endoplasm.

It has been shown that the myosin in amoebas can attach both to actin and to cell membranes. This ability suggests two possible mechanisms for extending the pseudopod. Myosin may form crossbridges between adjacent actin filaments and make them slide past each other, much as crossbridges in muscle slide thin filaments past thick ones. We saw in Chapter 1 how this process makes muscles shorten. Pseudopods may be made to extend by a similar process, working in reverse. Alternatively, the myosin may form bridges between actin filaments and the cell membrane and may make them move relative to each other. Either of these possibilities could be the basic mechanism of amoeboid crawling.

Yet another, very different mechanism has been suggested, and this mechanism does not involve the myosin at all. In a resting amoeba, all of the ectoplasm is presumably jellylike and filled with a network of actin filaments. Where a pseudopod is to form, the network must break down to allow the cell to bulge out. As the filaments break down into their constituent molecules, the concentration of molecule-sized particles in that part of the cell will be increased. Consequently, water will be drawn in by osmosis from other parts. The incoming water will make the cell membrane bulge out, forming a pseudopod which can be stiffened by allowing the network to re-form. Amoebas are traditionally thought of as the simplest of animals, yet our knowledge of their movement is so uncertain that we cannot decide between these very different theories.

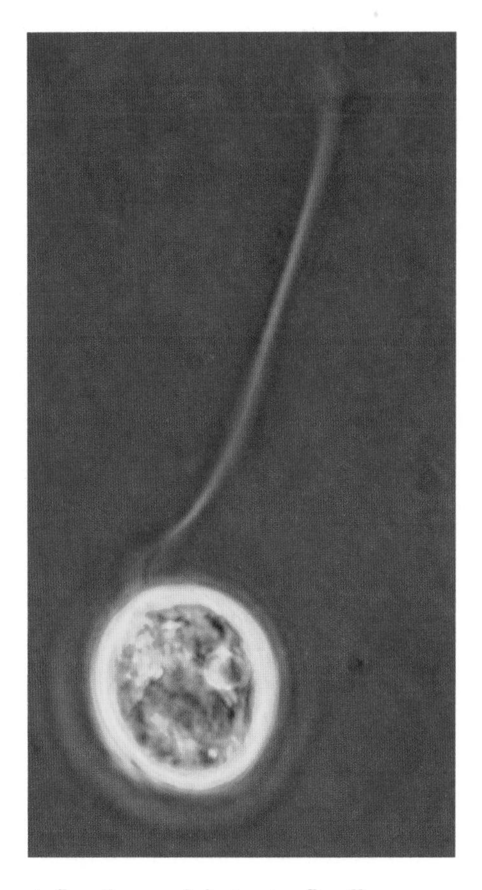

A flagellate undulating its flagellum to pull itself through the water. This is *Trachelomonas oblonga*.

## The Swimming of Flagellates

We have a much better understanding of the movement of another group of protistans, the flagellates. These tiny organisms are as important to life in the sea as green plants are to life on land. They are green like land plants, for they also contain chlorophyll and capture some of the sun's energy by photosynthesis, combining carbon dioxide

According to one possible mechanism for pseudopod formation, water would be drawn in osmotically when the actin filaments break down, extending the pseudopod.

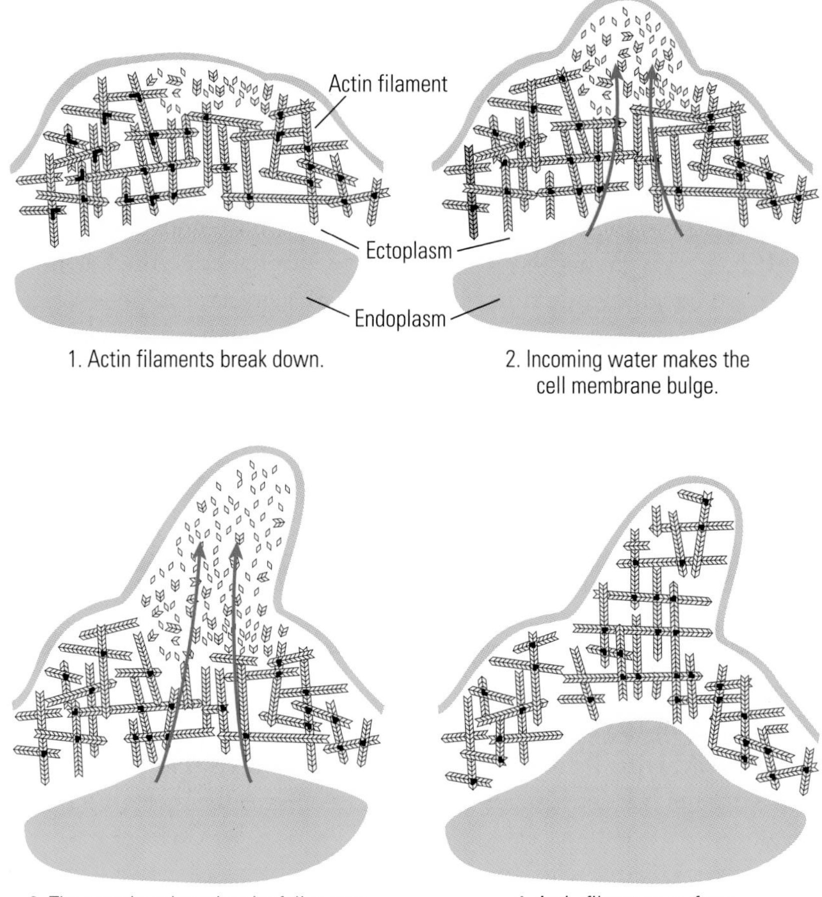


with water to form sugars. Just as life on land depends on the photosynthesis of plants, life in the sea depends on photosynthesis by flagellates and other green protistans that float as plankton in its upper layers.

Flagellates swim by beating their flagella, which are one, two, or occasionally more long whiplike projections from their bodies. Spermatozoa swim in the same way: their tails are flagella. Both flagellates and spermatozoa may seem to swim quite fast when they are watched through a microscope, but their speed is an illusion created by the magnification. The fastest swim only one fifth of a millimeter in one second, and the slowest manage only a millimeter in one minute.

Flagella are only just thick enough to be visible under the best light microscopes, since at one quarter of a micrometer the diameter of a flagellum is slightly less than the wavelength of light. They beat very fast, making as many as 70 cycles of movement per second, but their movements can be recorded either by high-speed moving photography or by multiple-flash still photography: in either case the camera must be attached to a high-quality microscope.

Flagellates swim in several different ways, but the flagellum usually undulates like the body of a swimming eel. Spermatozoa swim with the head in front, pushed along by the tail behind. Most flagellates swim with the flagellum in front, towing rather than pushing them.

Though the movements of spermatozoa look very like those of eels, there is an important mechanical difference. An eel or any other fish that suddenly stops making swimming movements will glide for-

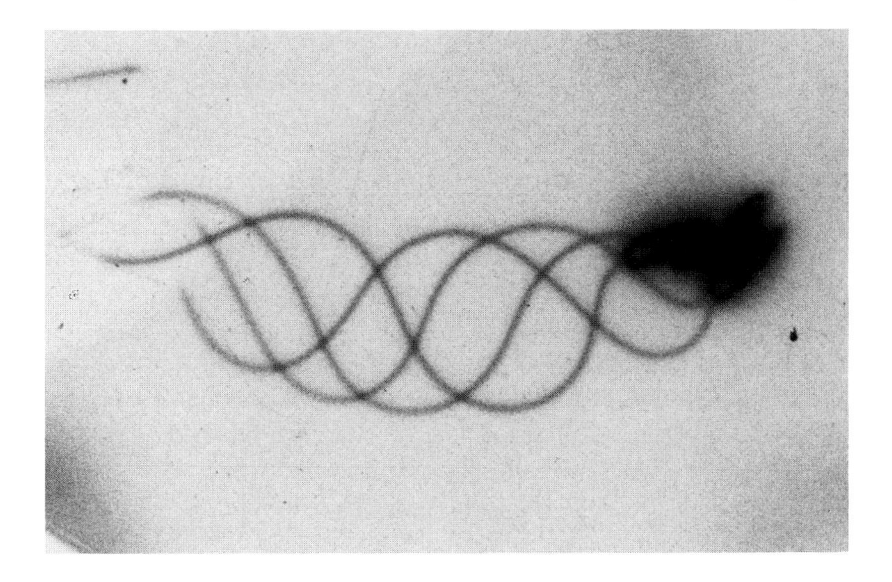

Four successive positions of the flagellum of a sea urchin sperm (*Lytechinus pictus*), captured by firing four flashes while the camera shutter was open.

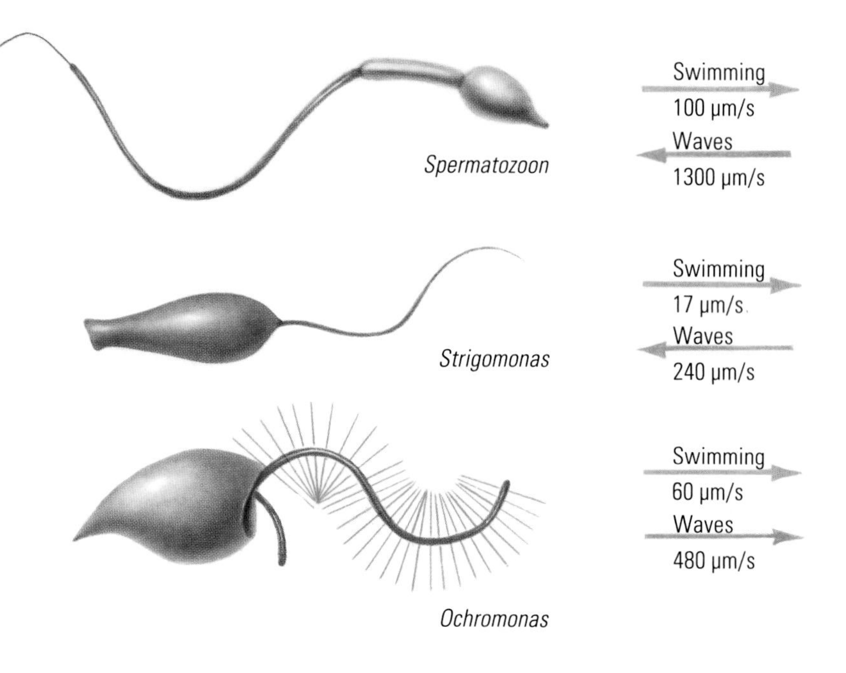


Flagella push spermatozoa from behind but pull the flagellates *Strigomonas* and *Ochromonas* from in front. The waves travel backward in the first two, but forward in *Ochromonas*.

ward for some distance, slowing down only gradually. In contrast, a spermatozoon that stops undulating its tail comes to an immediate halt. The reason for the difference is that spermatozoa are so small and so slow. Because they are slow, they have little momentum to carry them forward; and because they are small, they have a large ratio of surface area to volume, so the viscosity of the water slows them down very effectively. A shark that was shrunk to the size of a spermatozoon could still swim in water, but it would feel as if it were swimming in molasses. Sharks cannot stop suddenly, and many bony fishes spread their fins to slow themselves down, but spermatozoa have no need of brakes.

The dominant forces on swimming fishes and whales are inertial forces—that is, forces needed to accelerate the body or the water around it. In contrast, viscosity dominates the forces on small, slow swimmers such as spermatozoa. This difference between fishes and spermatozoa can be highlighted by calculating the Reynolds number, the equivalent for swimming of the Froude number that was so important in our discussion of running. Remember that

$$\text{Froude number} = \frac{(\text{speed})^2}{\text{length} \times \text{gravitational acceleration}}$$

It can be thought of as the ratio of inertial forces to gravitational forces. If gravity is important, different-sized bodies can move in dynamically similar fashion only if their Froude numbers are equal. Rather similarly, if viscosity is important, flow around different-sized bodies can be dynamically similar only if the bodies have equal Reynolds numbers:

$$\text{Reynolds number} = \frac{\text{speed} \times \text{length} \times \text{fluid density}}{\text{fluid viscosity}}$$

The Reynolds number represents the ratio of inertial forces to viscous forces.

Spermatozoa and eels are both long, thin swimmers, but the patterns of flow of water around them cannot be the same (the flows cannot be dynamically similar) unless their Reynolds numbers are equal. For water, density/viscosity is 1 million seconds per square meter. Thus, a spermatozoon 100 micrometers (0.0001 meter) long swimming at 100 micrometers per second (0.0001 meter per second) would have a Reynolds number equal to $0.0001 \times 0.0001 \times 1{,}000{,}000 = 0.01$. For a 1-meter eel swimming at 1 meter per second, the Reynolds number would be $1 \times 1 \times 1{,}000{,}000 = 1{,}000{,}000$. Consequently, the patterns of flow of water around swimming spermatozoa and eels are very different, although the patterns of movement of their undulating tails are rather similar.

To swim forward, most flagellates send waves backward along their flagella. The flagellate *Ochromonas* is a peculiar exception: it sends waves forward along its flagellum, but nevertheless travels forward. The reason became apparent when its flagellum was examined by electron microscopy and found to be fringed by an enormous number of exceedingly fine "flimmer filaments." In aggregate, these fila-

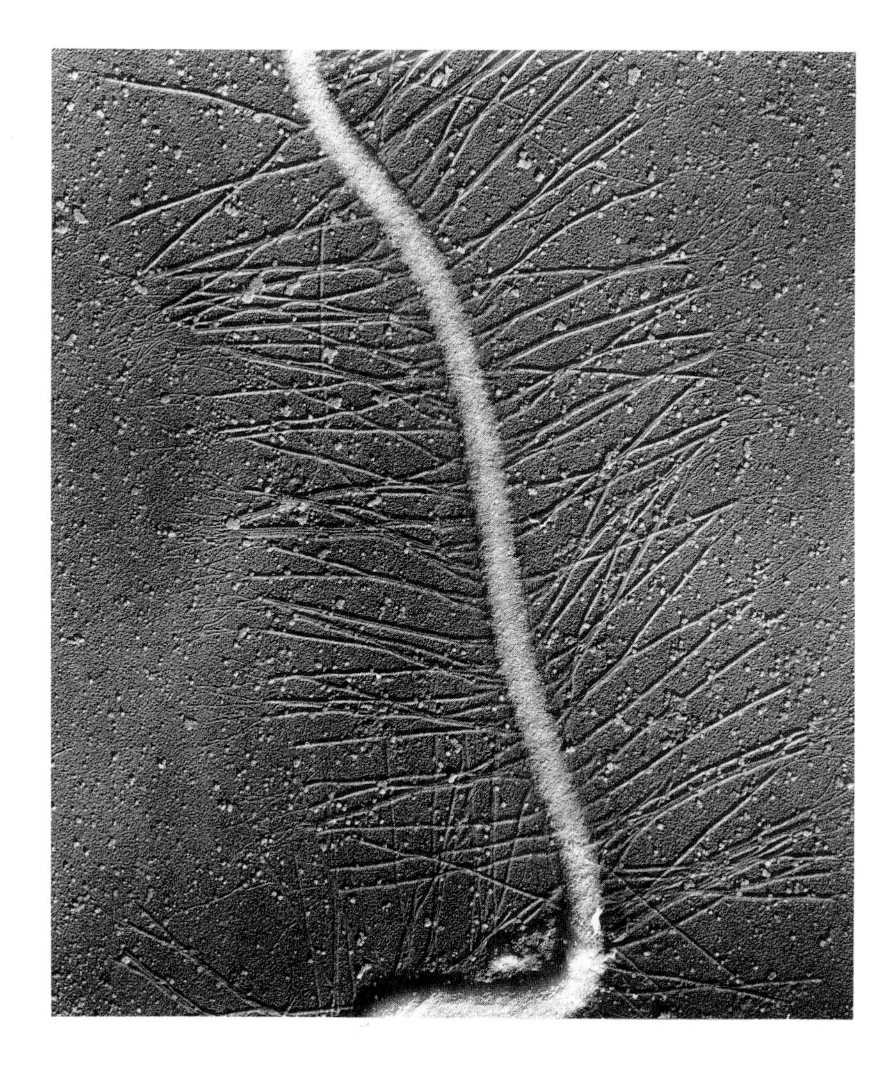

A flagellum fringed by flimmer filaments, seen in a transmission electron microscope.

ments are many times longer than the flagellum, so the hydrodynamic forces that act on them have much more effect on the organism's motion than do the forces on the main axis of the flagellum.

When we analyzed the serpentine crawling of snakes, we saw that waves traveling backward along the body drive the reptile forward, because the body can slide forward along its own track much more easily than it could slide broadside-on across the ground. Detailed analysis of the hydrodynamics of flagellate swimming shows that it works rather similarly. At low Reynolds numbers there is only half as much drag on a thin rod moving lengthwise through the water as there is on the same rod moving broadside-on at the same speed. The flimmer filaments lie in the plane of movement of the flagellum, at right angles to its main axis, so a movement that is lengthwise for the main axis is broadside-on for them, and vice-versa. For that reason, movements that would move an ordinary smooth flagellum backward drive one that has flimmer filaments forward.

*Chlamydomonas* and some other flagellates row with their flagella instead of undulating them. The flagella are held stiff during their backward power stroke so that they move broadside-on through the water and exert as much force as possible. Then, during the forward recovery stroke, they are allowed to bend so that they pull themselves lengthwise through the water and exert as little force as possible. The Reynolds number is too low for the organism's momentum to keep it moving forward during the recovery stroke, and it actually moves a little backward. The organism moves forward during the power stroke and a little backward during the recovery stroke, but the overall effect of each complete cycle of movement is to advance it through the water.

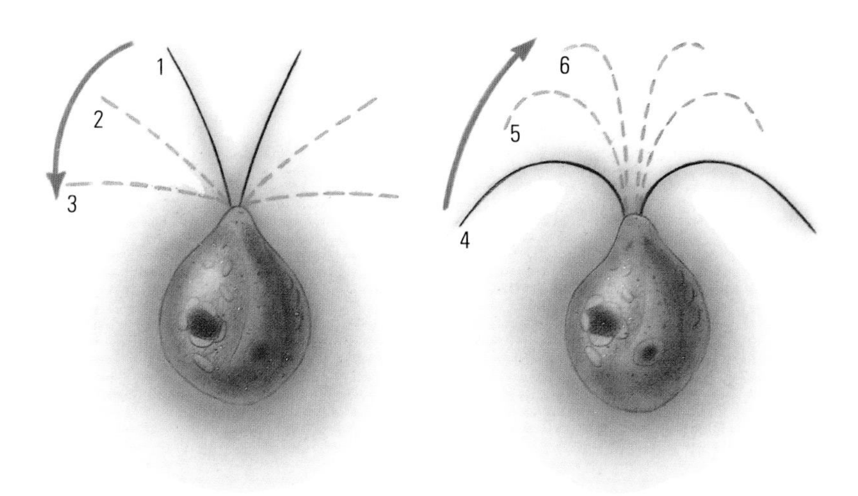


*Chlamydomonas* uses its two flagella to row itself along.

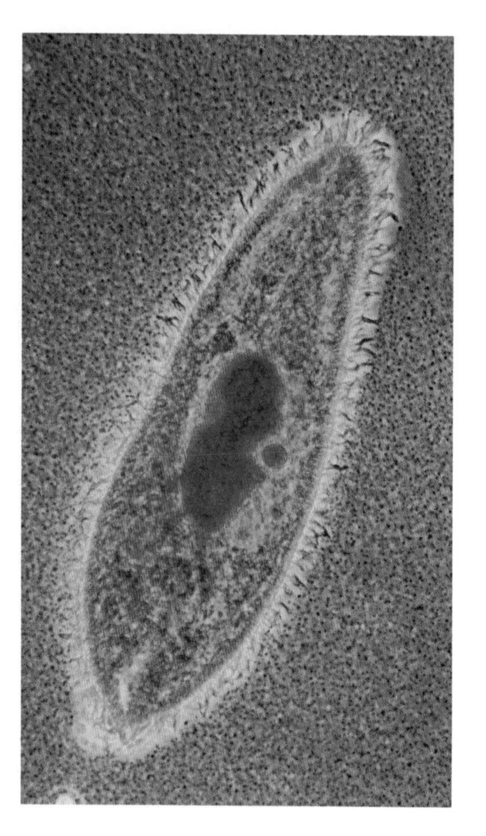

A ciliate protozoan *(Paramecium)*, about 0.2 millimeter long, seen by phase contrast microscopy, which makes it appear unnaturally colored. The whole body is covered by a carpet of cilia.

## Rowing with Cilia

The ciliates are another important group of microorganisms. Many of them are planktonic but (unlike most flagellates) they have no chlorophyll. They do not photosynthesize like plants but feed like animals, eating flagellates and other microorganisms. The motors that power them are similar to the flagella of flagellates in some ways, but very different in others.

Flagella are generally longer than the cell that bears them, and each cell has only one flagellum or a few. Ciliate microorganisms bear cilia instead: these are just like flagella except that they are much shorter (considerably shorter than the cell) and very much more numerous, often covering the entire body. Indeed, they are so much more numerous that ciliates have much bigger motors than flagellates of similar size. For example, the flagellate *Euglena* and the ciliate *Tetrahymena* are each about 50 micrometers ($\frac{1}{20}$ millimeter) long, but whereas *Euglena* has a single flagellum 50 micrometers long, *Tetrahymena* has about 500 cilia, each 7 micrometers long, giving a total length of 3500 micrometers. The cilia of *Tetrahymena* have the same diameter as the flagellum of *Euglena* and the same internal structure, so *Tetrahymena* has 70 times more material to drive it than *Euglena* has. Not surprisingly, ciliates swim much faster than flagellates. Whereas most flagellates swim at 0.02 to 0.2 millimeter per second, ciliates generally move at speeds between 0.4 and 2 millimeters per second.

Cilia make rowing movements rather like those of the flagella of *Chlamydomonas,* swinging stiffly back in their power stroke and bending much more in the recovery stroke. They are closely spaced and would collide if their beating were not coordinated, even if they all beat in the same direction. (Similarly, if I walked very close behind you, we would kick each other unless we kept in step.) Cilia actually beat with the precision of a regiment of drilling guardsmen; but, unlike soldiers, they do not all keep precisely in step. Different organisms use different patterns of beating, and I will describe just one, which is used by *Paramecium* and many other ciliates.

In these ciliates, the cilia are set out in a regular grid in lines running (we will suppose) north-south and east-west. Suppose that the direction of beating is north-south, with southward power strokes and northward recovery strokes. In that case the organism will be propelled toward the north. All the cilia in each north-south row will be found to be beating in phase with each other, but each of these rows beats slightly in advance of the next row toward the east. Waves of movement seem to travel from west to east like the ripples seen when wind blows over a field of wheat. The cilia seem to maintain coordination without any control being exerted by the organism; instead, this coordination seems to depend on hydrodynamic interaction between the cilia—that is, on the tendency of one cilium to drag its neighbors along with it. The east-west waves are a consequence of the cilia beating asymmetrically, bending a little toward the east in the recovery stroke.

The viscosity of the surrounding water has some important consequences for swimming with cilia. In a discussion of snail crawling in Chapter 3, I described a simple experiment to illustrate the phenomenon of viscosity. If molasses is sandwiched between two flat plates, its viscosity resists sliding movements of one plate over the other. The faster the movement, the greater the resisting force. I might have added that the thinner the layer of molasses, the greater the resistance to movement at the same speed.

A field of cilia beating backward and forward sets up the same sort of movement in the water as the sliding plates set up in the molasses. We can think of the cell surface as one plate, and we can imagine the other plate at the level of the tips of the cilia. The major part of the hydrodynamic force on a beating cilium is the force cre-

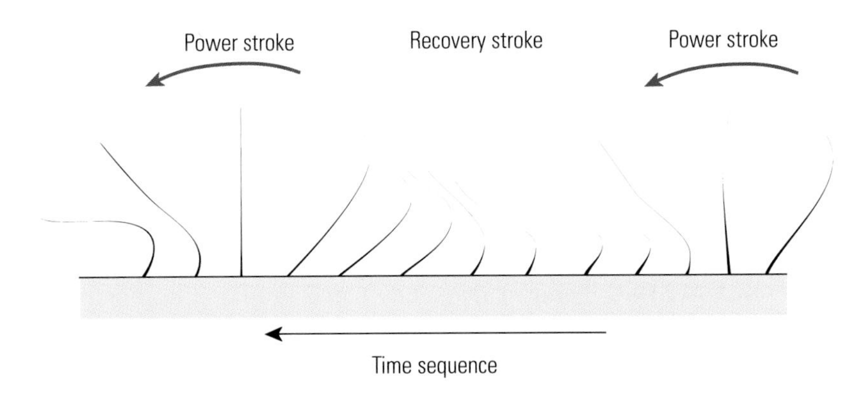


The rowing movements of a cilium.

ated by the viscosity of the thin layer of water between the cell surface and the tips of the cilia. This is the force the cilia must overcome to move the organism. Just as less force is needed to slide one plate over the other (at the same speed) if the molasses layer is thick, long cilia need not exert as much force (at the same swimming speed) as short ones.

You might conclude that the longer the cilia, the faster the organism should be able to swim, but this is true only up to a certain limit. The reason depends on the concept of bending moments. Suppose you are a rioter engaged in the dangerous sport of pulling down statues of discredited politicians. You are more likely to succeed if you attach your rope to a statue's neck than if you tie it to the ankles. The higher the rope is attached, the more leverage you will have about the base, and the more likely you are to be able to topple the statue. The quantity that matters in this context is the "bending moment" about the base: the force in the rope multiplied by the distance from the rope to the base of the statue. In ciliates, doubling the length of the cilium halves the force required (for the same swimming speed), but it also doubles the distance from the cilium base at which the force must act. The bending moment at the base of the cilium is the force multiplied by the distance, so halving the force and doubling the distance leaves it unchanged. The bending moment that the cilium can exert is limited, and consequently there is a limit to swimming speed that cannot be passed simply by making the cilia longer.

Although very long cilia cannot increase swimming speed, very short ones would reduce it. Remember from Chapter 1 that muscles cannot exert as much force when shortening fast as when shortening slowly. Similarly, cilia and flagella cannot exert bending moments as large when bending fast as when bending slowly. That faster bending movements are weaker was demonstrated by Kazuhiro Oiwa and Keiichi Takahashi of the University of Tokyo in an experiment of remarkable technical skill performed on the flagella of sea urchin spermatozoa. The shorter the cilium, the faster the base must bend to move the tip at the same speed, so very short cilia could not produce the bending moments required for swimming at normal speeds. Longer cilia need not bend so fast, but, since the bending moment cannot exceed a certain limit no matter how slow the bending, long cilia give no advantage in speed.

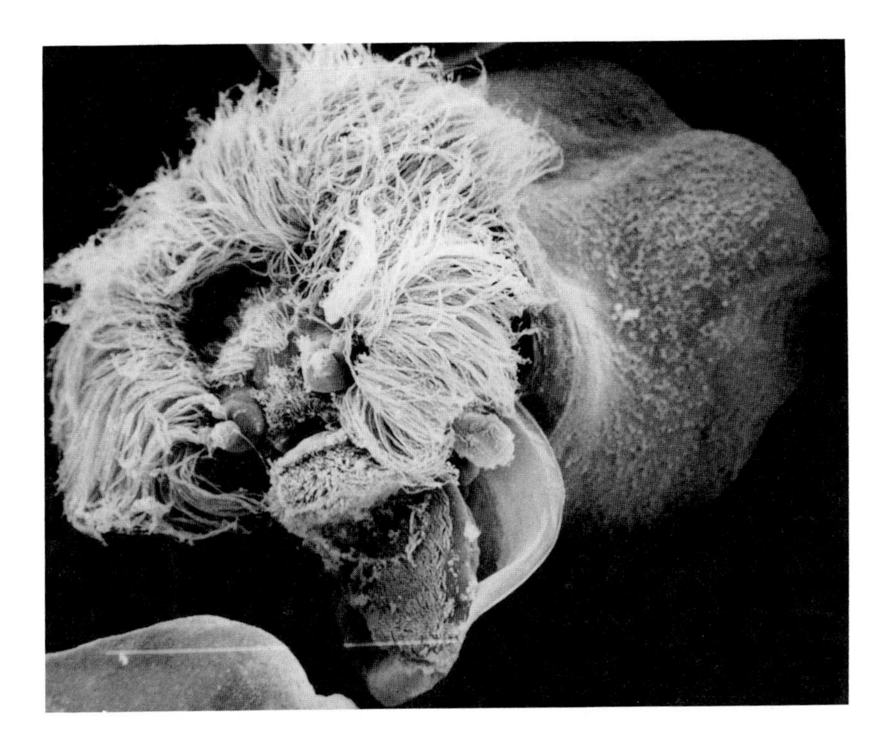

The planktonic larva of a mollusc (the abalone *Haliotis rufescens*) seen by scanning electron microscopy, showing the long cilia that power its swimming. Its diameter is 0.2 millimeter.

In addition to the ciliates, various many-celled animals of similar size swim by means of cilia. The pond-dwelling rotifers, which contain hundreds of cells (illustration on page 217) have already been mentioned. Many marine molluscs and worms hatch as tiny larvae that swim in the plankton for a while by means of cilia before settling on the bottom and developing their adult form. Unlike the single-celled ciliates that we have discussed so far, whose cilia cover the entire surface of their bodies, many-celled organisms may have only patches or bands of cilia. Rotifers, for example, have just a ring of cilia at the front end of the body. Many planktonic larvae have bands of cilia, and the rest of the body is bare.

Having bands of cilia instead of a complete covering can probably save energy. Most of the work required for swimming with cilia is needed to overcome the viscosity of the thin layer of water between the cell surface and the tips of the cilia. If this layer covers only part of the body, less work is needed.

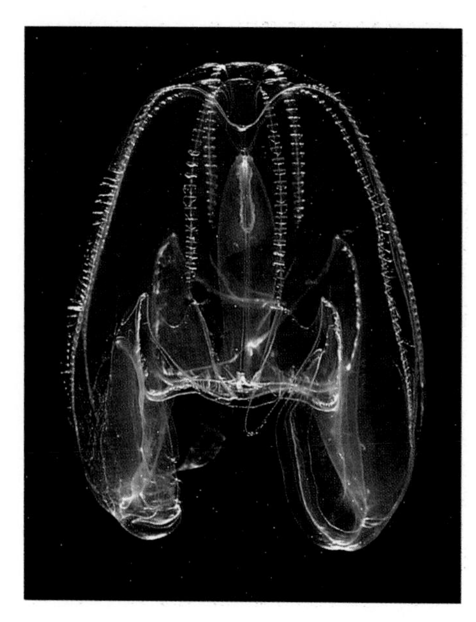

Rows of comb plates (groups of cilia) power the swimming of ctenophores such as this *Bolinopis chuni.* With diameters that are often more than a centimeter, ctenophores are much larger than other animals that use cilia for swimming.

Most of the many-celled animals that use cilia for swimming are no larger than the largest ciliates, 2 millimeters long or less. This is probably because ordinary cilia seem unable to move their possessors faster than about 2 millimeters per second, because the bending moments they can exert are limited. Larger animals can generally swim faster by using other means of propulsion. There is, however, one group of larger animals that swim faster by means of cilia, but these cilia are very remarkable. These animals are the ctenophores, gelatinous animals so transparent that they are difficult to see in water but extremely beautiful when you do see them. One example is the sea gooseberry *Pleurobrachia,* which has a diameter of about 15 millimeters and swims at about 8 millimeters per second. Instead of having separate cilia, it has comb plates, bundles of about 100,000 cilia that are packed very tightly together and beat as a unit. These cilia are up to 0.7 millimeter long, immensely longer than ordinary cilia, yet because there are so many in each comb plate, large bending moments can be exerted. The essential point about comb plates is that very large numbers of cilia are mounted on a very small area of the animal's surface.

## Inside a Bending Cilium

It's clear how the bending of cilia and flagella propels organisms through liquid, but what makes these fibers bend in the first place? One possibility is that they are passive whips, moved from the base by some motor in the body of the cell. That suggestion was disposed of by Ken Machin of Cambridge University when he showed by mathematical analysis that waves of the shape seen in photographs of swimming spermatozoa could not be produced by waving a passive structure from one end. Also, flagella broken off their cells by centrifugation will swim in suitable solutions, suggesting that the mechanism moving the flagella lies inside these processes.

Cilia and flagella are barely thick enough to be visible by light microscopy, so electron microscopy is essential to reveal their internal structure. Sections examined in this way show exceedingly thin tubular structures, called microtubules, running lengthwise along the cilium or flagellum: two single microtubules in the center are surrounded by a ring of nine double microtubules. This nine-plus-two pattern is exactly the same in cilia as in flagella. Short projections called dynein arms extend from each double microtubule toward the next one, and longer projections called radial spokes extend in toward the center.

With this structure, cilia and flagella could work in various ways. One possibility is that the double microtubules lengthen and shorten like muscle fibers: a flagellum would bend to the left if the microtubules along its left side shortened. Another is that the microtubules may remain constant in length but slide past each other, driven by the dynein arms in much the same way as the crossbridges of muscle slide the thin filaments past the thick ones. This second possibility was confirmed by Fred Warner and Peter Satir of Syracuse University, New York, and the University of California, Berkeley, who studied cilia from the gills of freshwater mussels. From electron micrographs of sections showing cilia cut longitudinally, they showed that the radial spokes on each double microtubule are arranged in groups of three. When the cilium is straight, groups of spokes from microtubules on opposite sides of the cilium are in register. When it

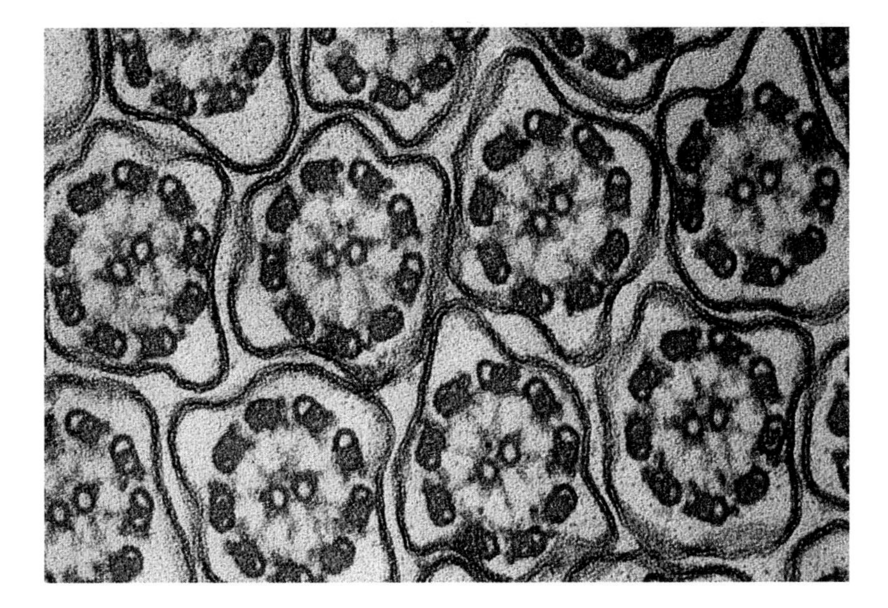

Closely packed cilia seen in an electron microscope picture. Each cilium has a diameter of 0.25 micrometer ($\frac{1}{4000}$ millimeter) and contains nine double microtubules encircling two single ones.

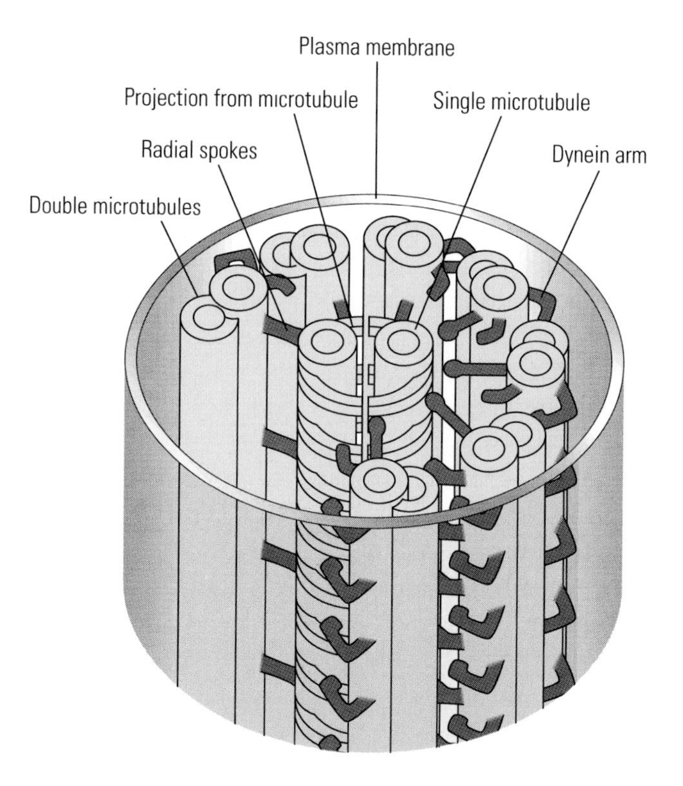


The internal structure of a cilium or flagellum.

bends, the spacing of the spokes remains unchanged (showing that the microtubules are not lengthening or shortening), but groups of spokes from microtubules on the inside of the bend are not always directly opposite those from microtubules on the outside. For example, in the diagram on this page the groups of spokes numbered 1 are in register but groups 2, 3, and upward are progressively more staggered so that group 7 on the inside of the bend comes into register with group 8 on the outside. The sliding of neighboring filaments past each other is presumably driven by the dynein arms.

## Sinking Down and Swimming Up

The flagellates and other green protistans that live in the sea, getting their energy by photosynthesis, must stay reasonably near the surface to survive. If they sink too deep, the light will be too dim for photosynthesis and they will perish. They can survive only down to 100 meters or so in the clear water of the oceans, and they must remain even nearer the surface in murky coastal waters.

These organisms face a challenge in remaining near the surface: because they are denser than seawater, commonly about 5 percent denser, they sink in still water. You might think that they must swim constantly upward to keep themselves in the brightly lit upper layer of the sea, where photosynthesis is possible. Unexpectedly, that is not necessarily the case.

The water in the sea is not still but turbulent, full of swirling eddies driven by wind, waves, and currents. In any particular interval of time a protistan is most likely to sink, but it could rise instead if it were moved by an eddy. As some cells are carried upward by eddies others will be hastened in their descent, and the average sinking rate will be the same as if the water were still. Given good luck, a protistan may meet a succession of eddies that will keep it near the surface for a long time, but it must eventually sink to the dark depths if it does not swim. Turbulence alone is not enough to maintain a population in the upper waters. Some other influence is needed, and that turns out to be reproduction.

Protistans generally reproduce by splitting into two halves, each of which becomes a half-sized protistan and then grows. If the cells

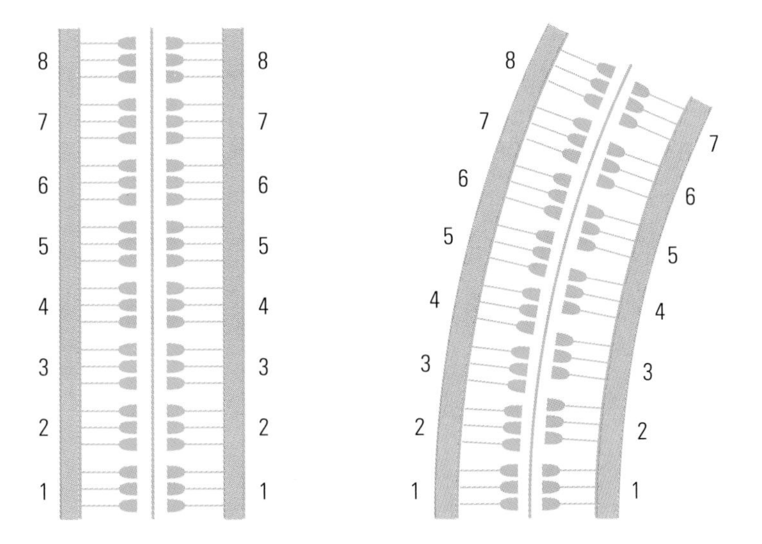


Groups of radial spokes on opposite sides of a cilium move out of register when it bends.

that are moved upward by eddies reproduce fast enough, they may be able to replace the ones lost by sinking. Turbulence and reproduction together may be able to maintain the population. The more slowly the cells sink and the faster they reproduce, the more likely a population will survive near the surface.

Small objects sink more slowly than large ones of equal density, both in air and in water. For example, raindrops fall fast, but fine droplets of mist sink to the ground very slowly. Pebbles sink rapidly through water, but fine silt particles settle out slowly. The smaller a planktonic organism is, the more slowly it is likely to sink.

Small organisms also breed fast. A population of bacteria may double in a few hours, but even in the most favorable conditions a population of whales would (unfortunately) need many years to do the same. Because small organisms sink slowly and breed fast, turbulence may be enough to maintain their populations even if they do not swim.

I have made a rough calculation to find the largest size an organism can be if a population is to remain near the surface without swimming, assuming a typical degree of turbulence in the sea. The maximum seems to be about 2 micrograms, about the mass of a moderately large ciliate. An organism of this mass would have a diameter of about 0.15 millimeter; it would probably sink in still water at a rate of about 0.6 millimeter per second; and its generation time would most likely be about 22 hours. Flagellates generally seem to be small enough to survive in the surface waters without swimming, despite being denser than the water, but most of the larger plankton that feed on them must swim.

Some planktonic protistans have no means of swimming and appear to stay high in the water by just these means. Yet although flagellates are small enough to do likewise, they can and do swim. Many of them (*Chlamydomonas* is an example) are bottom-heavy, denser at the hind end and less dense at the front end bearing the flagella. This means that when they swim in still water, gravity will tend to pull them into a bottom-down position, and their swimming action will move them upward. Swimming counteracts the tendency to sink, reducing the losses that would otherwise occur.

Swimming by flagellates gives rise to some curious effects that have been investigated by John Kessler of the University of Arizona and Tim Pedley of Leeds University. When Kessler lectures about

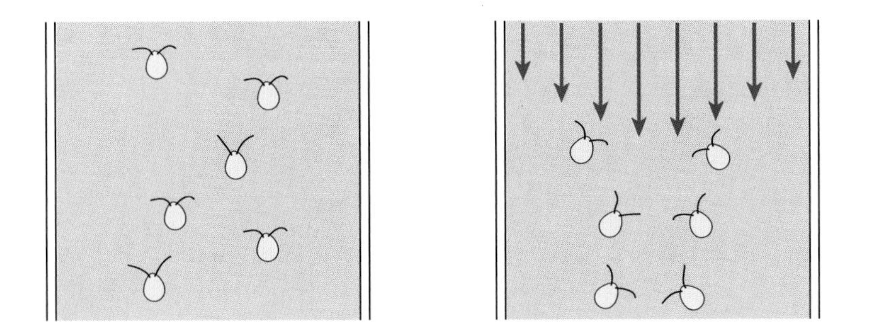

*Chlamydomonas* swim vertically upward in still water in a tube *(left)*. However, if the water is made to flow downward *(right)*, it rotates the organisms, making them swim toward the center.

these effects, he likes to show a very simple experiment. He sets up a vertical glass tube with a tap at the bottom so that he can allow liquid to flow slowly down through the tube. He takes water containing a high concentration of *Chlamydomonas* (their chlorophyll makes the suspension bright green) and stirs it well, then pours it into the top of the tube. As it flows down, the cells become concentrated in the tube's center, leaving clear water around a green core.

This phenomenon is called hydrodynamic focusing. It happens because fluid flows relatively slowly near the walls of a tube, where it is slowed down by viscosity, and fastest near the center. The bottom-heavy protistans would swim vertically upward if the water were still, but they are swimming against a current that is faster on one side of the cell (toward the center of the tube) and slower on the other (toward the wall). This uneven current swivels the cells, turning them toward the center of the tube so that their swimming moves them toward the center. Hydrodynamic focusing is the result of their swimming, of their being bottom-heavy, and of the faster flow in the center of the tube.

In another simple experiment, John Kessler fills a glass beaker with a suspension of *Chlamydomonas,* stirs it well, and leaves it sitting on a laboratory bench. At first the liquid is uniformly green, but within a few minutes vertical streaks of deeper green become distinct from the paler green of the rest of the suspension. The reason is that the upward swimming of the flagellates concentrates them near the surface. Because they are denser than water, the uppermost layer of the suspension becomes denser than the rest, an unstable situation.

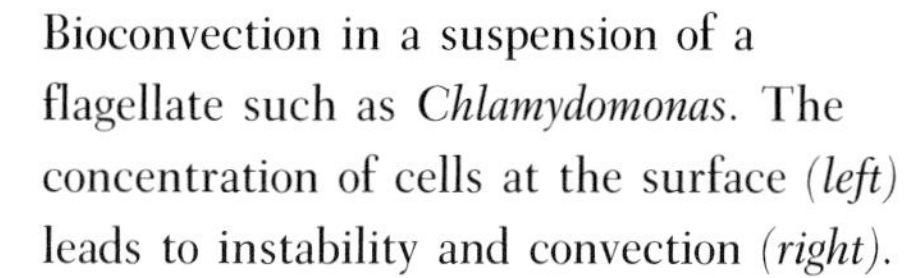

Bioconvection in a suspension of a flagellate such as *Chlamydomonas*. The concentration of cells at the surface *(left)* leads to instability and convection *(right)*.

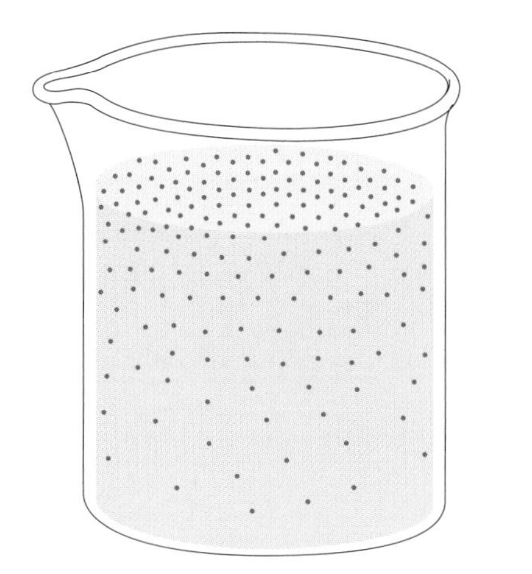

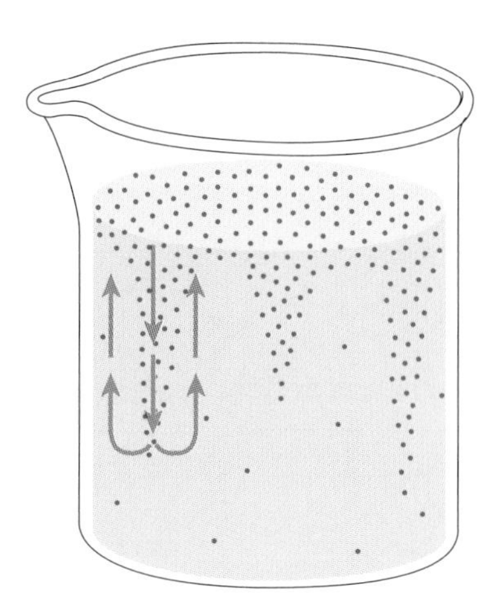

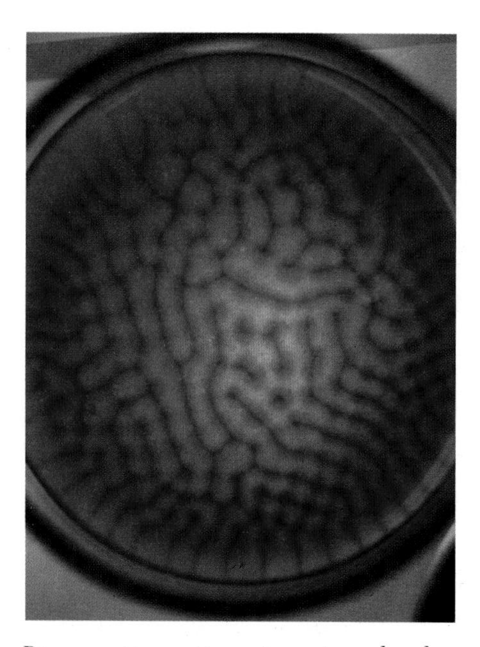

Bioconvection patterns in water colored green by a dense population of a microscopic alga *(Dunaliella tertiolecta)*. The diameter of the dish is 55 millimeters and the water is about 3 millimeters deep.

Parts of the dense suspension start sinking from the surface, and as they sink, hydrodynamic focusing draws in nearby cells. As more cells are drawn into the sinking regions, the regions become even denser and sink faster. The result is a pattern of dense, deep-green plumes sinking through the much paler bulk of the suspension. The pale parts must rise to make room for the sinking plumes, and the water is set circulating in a pattern very much like the convection currents in a heated saucepan. Indeed, the phenomenon is called bioconvection. In shallow dishes, bioconvection can set up strikingly regular patterns, as an illustration shows.

The tiny organisms that we have been discussing in this chapter move without muscles. Amoebas crawl by forming pseudopods and flowing into them. Flagellate protozoans swim by undulating their long flagella or by moving them like oars. Ciliate protozoans are driven by large numbers of cilia, which are much shorter than flagella but are powered by the same internal mechanism. All these organisms can swim only slowly and so are at the mercy of water currents and turbulence, but some tiny plankton that do not swim depend on turbulence for their very survival: it stirs up the water so that they do not all sink to the dark depths where they could not photosynthesize. Finally, we have seen how the swimming of green flagellates can make them clump together in strange patterns, by the process of bioconvection.

# Epilog: What We Want to Know Next

This book has shown that we know a great deal about animal locomotion, but I do not want to leave the impression that we know and understand even most of what there is to know. It would be tedious to try to list all the things that have not been adequately explained; instead I will use just two examples to illustrate our ignorance.

First, there is the very basic problem of explaining the energy consumption of muscles. I explained in Chapter 1 that not only do muscles use metabolic energy whenever they do work, but they also use energy without doing work whenever they exert forces without shortening. I did not explain properly how the metabolic energy that is needed for particular activities is used: the reason is that neither I nor (I think) anyone else understands it properly. You might reasonably expect that if you told us precisely what a muscle does, how much it lengthens and shortens, and what forces it exerts, we would be able to calculate the metabolic energy it would use, but we simply cannot. This difficulty became particularly obvious in Chapter 2, where I argued that the elasticity of tendons saves energy in running, but was confusingly vague about how it does this. I pointed out that it reduces the work that the muscles have to do, but then went on to say that it enables us to make do with shorter or slower muscle fibers than we would otherwise need, and these shorter or slower muscle fibers can be expected to develop force more economically. I was keeping my options open because neither I nor anyone else is clear about the relative importance of the energy cost of doing work and that of exerting forces. My colleagues and I were the first to show that tendons rather than muscles are the important springs in the legs of mammals, so you might reasonably expect us to be clear about these things; but we are not.

The other problem that I want to highlight concerns flight. Biologists used to try to explain how birds and insects fly, by using the aerodynamics that engineers use when dealing with fixed-wing aircraft. During the past 20 years it has become progressively clearer that fixed-wing aerodynamics is not adequate, because the aerodynamics of flapping flight is complicated by the unsteady effects that arise whenever a wing comes to a halt at the end of a stroke and starts moving in the opposite direction. Great progress has been made since the importance of unsteady effects was first recognized. Torkel Weis-Fogh explained the clap and fling mechanism (illustrated on page 125); Jeremy Rayner and Charlie Ellington developed theories of flight that did not use the aerodynamics of fixed-wing aircraft but looked directly at the air movements that result from the movements of flight; and the Bristol group showed us the patterns of air movement behind flying birds and bats (illustrated on pages 129 and 130). Despite all this, we still cannot in most cases explain quantitatively how particular unsteady movements of wings set up particular patterns of movements in the air. I do not think that anyone can as yet calculate confidently the aerodynamic forces on the wings of a dragonfly or a hoverfly and show how they balance the animal's weight.

Like most insects, dragonflies have two pairs of wings, but whereas other insects beat the pairs in synchrony, dragonflies beat them alternately. This is *Aeschna cyanea*.

There are many other problems like these that we know we have not solved. More insidious, perhaps, are the others that we think we have solved but have actually misunderstood. We thought we knew how slugs and snails crawl, for example, until Mark Denny showed us that the supposed explanation could not work and proposed another (which I explained in Chapter 3).

# Further Reading

## General


Alexander, R. McN. (1982). *Locomotion of Animals*. Glasgow: Blackie.

Gray, J. (1968). *Animal Locomotion*. London: Weidenfeld & Nicolson.

McMahon, T. A., and J. T. Bonner (1983). *On Size and Life*. New York: Scientific American Library.


## Chapter 1


Heglund, N. C., and G. A. Cavagna (1987). Mechanical work, oxygen consumption and efficiency in isolated frog and rat striated muscle. *American Journal of Physiology* 253: C22–C29.

McMahon, T. A. (1984). *Muscles, Reflexes and Locomotion*. Princeton, N.J.: Princeton University Press.

Woledge, R. C., N. A. Curtin, and E. Homsher (1985). *Energetic Aspects of Muscle Contraction*. London: Academic Press.


## Chapter 2


Alexander, R. McN. (1984). Walking and running. *American Scientist* 72: 348–354.

Alexander, R. McN. (1988). Why mammals gallop. *American Zoologist* 28: 237–245.

Alexander, R. McN. (1984). *Elastic Mechanisms in Animal Movement*. Cambridge: Cambridge University Press.

Alexander, R. McN. (1989). Optimization and gaits in the locomotion of vertebrates. *Physiological Reviews* 69: 1199–1227.

Alexander, R. McN. (1992). *The Human Machine*. London: Natural History Museum Publications.

Alexander, R. McN., and A. S. Jayes (1983). A dynamic similarity hypothesis for the gaits of quadrupedal mammals. *Journal of Zoology* 201: 135–152.

Biewener, A. A. (1989). Scaling body support in mammals: limb posture and muscle mechanics. *Science* 245: 45–48.

Carrier, D. R. (1987). Lung ventilation during walking and running in four species of lizards. *Experimental Biology* 47: 33–42.

Hoyt, D. F., and C. R. Taylor (1981). Gait and the energetics of locomotion in horses. *Nature* 292: 239–240.

Jayes, A. S., and R. McN. Alexander (1980). The gaits of chelonians: walking techniques for very low speeds. *Journal of Zoology* 191: 353–378.

Ker, R. F., M. B. Bennett, S. R. Bibby, R. C. Kester, and R. McN. Alexander (1987). The spring in the arch of the human foot. *Nature* 325: 147–149.

Kram, R, and C. R. Taylor (1990). Energetics of running: a new perspective. *Nature* 346: 265–267.

Manton, S. M. (1965). The evolution of arthropodan locomotory mechanisms. Part 8. Functional requirements and body design in Chilopoda. *Journal of the Linnean Society (Zoology)* 45: 251–484.

Taylor, C. R., N. C. Heglund, and G. M. O. Maloiy (1982). Energetics and mechanics of terrestrial locomotion. I. Metabolic energy consumption as a function of speed and body size in birds and mammals. *Journal of Experimental Biology* 97: 1–21.


## Chapter 3


Bennet-Clark, H. C. (1975). The energetics of the jump of the locust, *Schistocerca gregaria*. *Journal of Experimental Biology* 63: 53–83.

Bennet-Clark, H. C., and E. C. A. Lucey (1967). The jump of the flea: a study of the energetics and a model of the mechanism. *Journal of Experimental Biology* 47: 59–76.

Denny, M. (1980). The role of gastropod pedal mucus in locomotion. *Nature* 285: 160–161.

Dixon, A. F. G., P. C. Croghan, and R. P. Gowing (1990). The mechanism by which aphids adhere to smooth surfaces. *Journal of Experimental Biology* 152: 243–253.

Evans, M. E. G. (1972). The jump of the click beetle (Coleoptera: Elateridae)—a preliminary study. *Journal of Zoology* 167: 319–336.

Fleagle, J. G. (1974). The dynamics of a brachiating siamang (*Hylobates (Symphalangus) syndactylus*). *Nature* 248: 259–260.

Gans, C. (1962). Terrestrial locomotion without limbs. *American Zoologist* 2: 167–182.

Hill, A. V. (1950). The dimensions of animals and their muscular dynamics. *Science Progress* 38: 209–230.

Jenkins, F. A., and D. M. McClearn (1984). Mechanisms of hind foot reversal in climbing mammals. *Journal of Morphology* 182: 197–219.

Stork, N. E. (1980). Experimental analysis of adhesion of *Chrysolina polita* (Chrysomelidae: Coleoptera) on a variety of surfaces. *Journal of Experimental Biology* 88: 91–107.

Trueman, E. R. (1975). *The Locomotion of Soft-Bodied Animals*. London: Arnold.

Walton, M., B. C. Jayne, and A. F. Bennett (1990). The energetic cost of limbless locomotion. *Science* 249: 524–527.


## Chapter 4


Gibo, D. L., and M. J. Pallett (1979). Soaring flight of monarch butterflies, *Danaus plexippus* (Lepidoptera: Danaidae), during the late summer migration in southern Ontario. *Canadian Journal of Zoology* 57: 1393–1401.

Norberg, U. M. (1975). Hovering flight in the pied flycatcher *(Ficedula hypoleuca)*. In T. Y.-T. Wu, C. J. Brokaw, and C. Brennen (eds), *Swimming and Flying in Nature* 2: 869–881. New York: Plenum.

Pennycuick, C. J. (1968). A wind-tunnel study of gliding flight in the pigeon *Columba livia*. *Journal of Experimental Biology* 49: 509–526.

Pennycuick, C. J. (1972). Soaring behaviour and performance of some East African birds, observed from a motor glider. *Ibis* 114: 178–218.

Pennycuick, C. J. (1982). The flight of petrels and albatrosses (Procellariiformes), observed in South Georgia and its vicinity. *Philosophical Transactions of the Royal Society* B 300: 75–106.

Pennycuick, C. J. (1983). Thermal soaring compared in three dissimilar tropical bird species, *Fregata magnificens, Pelicanus occidentalis* and *Coragyps atratus*. *Journal of Experimental Biology* 102: 307–325.

Scholey, K. (1986). The climbing and gliding locomotion of the giant red flying squirrel *Petaurista petaurista* (Sciuridae). In W. Nachtigall (ed.), *Biona Report* 5: *Bat flight—Fledermausflug,* 187–204. Heidelberg: Fischer.

Videler, J., and A. Groenewold (1991). Field measurements of hanging flight aerodynamics in the kestrel *Falco tinnunculus*. *Journal of Experimental Biology* 155: 519–530.

Withers, P. C. (1979). Aerodynamics and hydrodynamics of the "hovering" flight of Wilson's storm petrel. *Journal of Experimental Biology* 80: 83–91.


## Chapter 5


Dudley, R., and C. P. Ellington (1990). Mechanics of forward flight in bumblebees (two papers). *Journal of Experimental Biology* 148: 19–88.

Ellington, C. P. (1984). The aerodynamics of hovering insect flight. III. Kinematics. *Philosophical Transactions of the Royal Society* B 305: 41–78.

Ellington, C. P., K. E. Machin, and T. M. Casey (1990). Oxygen consumption of bumblebees in forward flight. *Nature* 347: 472–473.

Jenkins, F. A., K. P. Dial, and G. E. Goslow (1988). A cineradiographic analysis of bird flight: the wishbone in starlings is a spring. *Science* 241: 1495–1498.

Kokshaysky, N. V. (1979). Tracing the wake of a flying bird. *Nature* 279: 146–148.

Machin, K. E., and J. W. S. Pringle (1959). The physiology of insect fibrillar muscle. II. Mechanical properties of a beetle flight muscle. *Proceedings of the Royal Society* B 151: 204–225.

Norberg, U. M. (1989). *Vertebrate Flight: Mechanics, Physiology, Morphology, Ecology and Evolution*. Berlin: Springer.

Rayner, J. M. V. (1979). A new approach to animal flight mechanics. *Journal of Experimental Biology* 80: 17–54.

Rayner, J. M. V. (1985). Bounding and undulating flight in birds. *Journal of Theoretical Biology* 117: 47–77.

Rayner, J. M. V. (1987). Form and function in avian flight. *Current Ornithology* 5: 1–66.

Rayner, J. M. V., G. Jones, and A. Thomas (1986). Vortex flow visualizations reveal change in upstroke function with flight speed in bats. *Nature* 321: 162–164.

Spedding, G. R. (1987). The wake of a kestrel *(Falco tinnunculus)* in flapping flight. *Journal of Experimental Biology* 127: 59–78.

Spedding, G. R., J. M. V. Rayner, and C. J. Pennycuick (1984). Momentum and energy in the wake of a pigeon *(Columba livia)* in slow flight. *Journal of Experimental Biology* 111: 81–102.

Tucker, V. A. (1968). Respiratory exchange and evaporative water loss in the flying budgerigar. *Journal of Experimental Biology* 48: 67–87.

Wootton, R. J. (1990). The mechanical design of insect wings. *Scientific American* 263(5): 66–72.


## Chapter 6


Alexander, R. McN. (1965). The lift produced by the heterocercal tails of Selachii. *Journal of Experimental Biology* 43: 131–138.

Alexander, R. McN. (1989). Size, speed and buoyancy adaptations in aquatic animals. *American Zoologist* 30: 189–196.

Berg, T., and J. B. Steen (1968). The mechanism of oxygen concentration in the swimbladder of the eel. *Journal of Physiology* 195: 631–638.

Bidigare, R. R., and D. C. Biggs (1980). The role of sulfate exclusion in buoyancy maintenance by siphonophores and other gelatinous zooplankton. *Comparative Biochemistry and Physiology* 66A: 467–471.

Corner, E. D. S., E. J. Denton, and G. R. Forster (1969). On the buoyancy of some deep sea sharks. *Proceedings of the Royal Society* B 171: 415–429.

Denton, E. J. (1974). On buoyancy and the lives of fossil and modern cephalopods. *Proceedings of the Royal Society* B 185: 273–299.

Jones, F. R. H., and P. Scholes (1985). Gas secretion and resorption in the swimbladder of the cod, *Gadus morhua*. *Journal of Comparative Physiology* B 155: 319–331.

## Chapter 7

Alexander, R. McN. (1967). *Functional Design in Fishes*. London: Hutchinson.

Altringham, J. D., and I. A. Johnston (1990). Modelling muscle power output in a swimming fish. *Journal of Experimental Biology* 148: 395–402.

Au, D., and D. Weihs (1980). At high speeds dolphins save energy by leaping. *Nature* 284: 548–550.

Baudinette, R. V., and P. Gill (1985). The energetics of "flying" and "paddling" in water: locomotion in penguins and ducks. *Journal of Comparative Physiology* B 155: 373–380.

Blake, R. W. (1983). *Fish Locomotion*. Cambridge: Cambridge University Press.

Brett, J. R. (1965). The swimming energetics of salmon. *Scientific American* 213(2): 80–85.

Gray, J. (1933). The movement of fish with special reference to the eel. *Journal of Experimental Biology* 10: 88–104.

Hui, C. A. (1987). The porpoising of penguins: an energy-conserving behaviour for respiratory ventilation? *Canadian Journal of Zoology* 65: 209–211.

Hui, C. A. (1988). Penguin swimming (two papers). *Physiological Zoology* 61: 333–350.

Lang, T. G. (1975). Speed, power and drag measurements of dolphins and porpoises. In T. Y.-T. Wu, C. J. Brokaw, and C. Brennen (eds), *Swimming and Flying in Nature* 2: 553–572. New York: Plenum.

Lighthill, M. J. (1969). Hydrodynamics of aquatic animal propulsion. *Annual Review of Fluid Mechanics* 1: 413–446.

McCutchen, C. W. (1977). Froude propulsive efficiency of a small fish, measured by wake visualization. In T. J. Pedley (ed.), *Scale Effects in Animal Locomotion*, 339–363. London: Academic Press.

Nachtigall, W. (1980). Mechanics of swimming in water beetles. In H. Y. Elder and E. R. Trueman (eds), *Aspects of Animal Movement*, 107–124. Cambridge: Cambridge University Press.

Prange, H. D., and K. Schmidt-Nielsen (1970). The metabolic cost of swimming in ducks. *Journal of Experimental Biology* 53: 763–778.

Rome, L. C., R. P. Funke, R. McN. Alexander, G. Lutz, H. Aldridge, F. Scott, and M. Freadman (1988). Why animals have different muscle fibre types. *Nature* 335: 824–827.

Webb, P. W. (1971). The swimming energetics of trout (two papers). *Journal of Experimental Biology* 55: 489–540.

Webb, P. W., and D. Weihs, (eds) *Fish Biomechanics*. New York: Praeger.

Webber, D. M., and R. K. O'Dor (1986). Monitoring the metabolic rate and activity of free-swimming squid with telemetered jet pressure. *Journal of Experimental Biology* 126: 205–224.

## Chapter 8

Adams, R. J., and T. D. Pollard (1989). Binding of Myosin I to membrane lipids. *Nature* 340: 565–568.

Alexander, R. McN. (1989). Size, speed and buoyancy adaptations in aquatic animals. *American Zoologist* 30: 189–196.

Amos, W. B., and J. G. Duckett, (eds) (1982). Prokaryotic and eukaryotic flagella. *Symposia of the Society for Experimental Biology* 35: 1–632.

Blake, J. R., and M. A. Sleigh (1974). Mechanics of ciliary locomotion. *Biological Reviews* 49: 85–125.

Chia, F.-S., J. Buckland-Nicks, and C. M. Young (1984). Locomotion of marine invertebrate larvae: a review. *Canadian Journal of Zoology* 62: 1205–1222.

Kessler, J. O. (1985). Hydrodynamic focusing of motile algal cells. *Nature* 313: 218–220.

Oiwa, K., and K. Takahashi (1988). The force-velocity relationship for microtubule sliding in demembranated sperm flagella of the sea urchin. *Cell Structure and Function* 13: 193–206.

Stossel, T. P. (1990). How cells crawl. *American Scientist* 78: 408–423.

Warner, F. D., and P. Satir (1974). The structural basis of ciliary bend formation: radial spoke positional changes accompanying microtubule sliding. *Journal of Cell Biology* 63: 35–63.

# Sources of Illustrations

Animal illustrations by Paul Mirocha; line illustrations by Fine Line Illustrations, Inc.

**Chapter 1** *Opposite p. 1:* M. P. Kahl/DRK Photo. *p. 3:* Edward Ross. *p. 4:* Alison Cutts. *p. 5:* M. Abbey/Photo Researchers. *p. 8:* After R. McN. Alexander, *Locomotion of Animals*, Blackie and Son Ltd, 1982. *p. 10:* Stephen Dalton/NHPA. *p. 14:* G. W. Willis/BPS. *p. 15:* Travis Amos.

**Chapter 2** *p. 16:* Rick Rickman/Duomo. *p. 19:* The International Museum of Photography at George Eastman House. *p. 21:* Gerry Ellis/The Wildlife Collection. *p. 22:* The International Museum of Photography at George Eastman House. *p. 24:* (*top*) Paul J. Sutton/Duomo; (*bottom*) N. C. Heglund. *p. 25:* After P. O. Astrand and K. Rodahl, *Textbook of Work Physiology, 3d ed.*, reproduced with permission of McGraw Hill, Inc., 1986. *p. 26:* After R. McN. Alexander, *Proceedings of the Royal Institution* 62:1–14, 1990. *p. 27:* Robert Knauft/Photo Researchers. *p. 30:* After R. McN. Alexander et al., *Journal of Zoology* 198:293–313, 1982. *p. 31:* After R. McN. Alexander, *Proceedings of the Royal Institution* 62:1–14, 1990. *p. 33:* (*top and middle*) The International Museum of Photography at George Eastman House; (*bottom*) Stanford University Museum of Art, 41.1018, Muybridge Collection. *p. 34:* The International Museum of Photography at George Eastman House. *p. 35:* After D. F. Hoyt and C. R. Taylor, *Nature* 292:239–240, 1981. *p. 37:* From L. S. Gambaryan, *How Mammals Run*, Wiley, 1974. *p. 41:* After F. A. Jenkins (Ed.), *Primate Locomotion*, Academic Press, 1974. *p. 43:* R. C. Taylor, Harvard University. *p. 44:* After C. R. Taylor, N. C. Heglund, and G. M. O. Maloiy, *Journal of Experimental Biology* 97:1–21, 1982. *p. 48:* (*left*) Frans Lanting/Minden Pictures; (*right*) K. Ammann/Planet Earth Pictures. *p. 50:* Art Wolfe. *p. 54:* After S. M. Manton, *Journal of Linnean Society (Zoology)* 45:251–284. *p. 55:* Frans Lanting/Minden Pictures.

**Chapter 3** *p. 56:* Stephen Dalton/NHPA. *p. 59:* After C. F. Oxnard, R. H. Crompton, and S. S. Lieberman, *Animal Lifestyles and Anatomies*, University of Washington Press, 1990. *p. 60:* Stephen Dalton/NHPA. *p. 62:* Dwight Kuhn. *p. 63:* (*left*) Stephen Dalton/NHPA; (*right*) After H. C. Bennett Clark, *Journal of Experimental Biology* 63:53–83, 1975. *p. 64:* After M. E. G. Evans, *Journal of Zoology* 167:319–336, 1972. *p. 65:* After J. Fleagle, *Nature*, 248:259–260, 1974. *p. 66:* Jean-Paul Ferrero/ AUSCAPE International. *p. 67:* K. Ghani/NHPA. *p. 68:* Drawing based on x-ray photo by F. A. Jenkins and D. M. McClearn, *Journal of Morphology*, 182:197–219, 1984. *p. 69:* H. Charles/Planet Earth Pictures. *p. 70:* Tim Laman/The Wildlife Collection. *p. 72:* (*left*) Nigel Stork, Natural History Museum; (*upper right*) Esther Beaton/AUSCAPE International; (*right*) E. E. Williams and J. A. Peterson, *Science* 215:1509–1511, 1982. *p. 73:* (*left*) Art Wolfe; (*right*) After R. McN. Alexander, *Locomotion of Animals*. *p. 74:* K. H. Switak/NHPA. *p. 76:* (*top*) Gary R. Zahm, DRK Photo; (*bottom*) After R. McN. Alexander, *Locomotion of Animals*. *p. 79:* Stephen Dalton/NHPA. *pp. 80 and 84:* After R. McN. Alexander, *Locomotion of Animals*.

**Chapter 4** *p. 86:* Nick Bergkessel. *p. 89:* Hellio and Van Igen/NHPA. *p. 96:* After R. McN. Alexander, *Locomotion of Animals*. *p. 99:* Sue Earle/Planet Earth Pictures. *p. 101:* Graham Robertson/AUSCAPE International. *p. 103:* B. K. Wheeler/VIREO. *p. 107:* Richard Coomber/Planet Earth Pictures. *p. 110:* Art Wolfe. *p. 115:* Jean-Paul Ferrero/AUSCAPE International. *p. 116:* From A. Charig, *A New Look at Dinosaurs*, British Museum (Natural History), 1979. *p. 117:* Frans Lanting/Minden Pictures.

**Chapter 5** *p. 118:* Art Wolfe. *p. 121:* Robert Tyrell. *p. 122:* (*top*) Stephen Dalton/NHPA; (*bottom*) Patricia J. Wynne and David J. S. Newman, University of Exeter. *p. 123:* After G. Ruppell, *Vogelflug*, Rowohlt Taschenbuch, Hamburg, Germany, 1980. *pp. 124 and 125:* After R. McN. Alexander, *Locomotion of Animals*. *p. 127:* Nikolai Kokshaysky, Moscow. *p. 128:* J. M. V. Rayner and A. L. R. Thomas. *pp. 129, 130 and 131:* After R. McN. Alexander, *Animals*, Cambridge University Press, 1990. *p. 132:* After F. A. Jenkins, *Science* 241:1495–1498, copyright 1988, American Association for the Advancement of Science.

*p. 134 and 135:* Art Wolfe. *p. 137:* Russ Hansen. *p. 138:* After R. McN. Alexander, *Locomotion of Animals. p. 139:* Art Wolfe. *p. 141:* After R. Dudley and C. P. Ellington, *Journal of Experimental Biology* 148:53–88, 1990. *p. 143:* Stephen Dalton/NHPA. *p. 145:* After J. M. V. Rayner, *Current Ornithology* 5:1–77. *p. 151:* After R. McN. Alexander, *Locomotion of Animals. p. 152:* Gerry Ellis/The Wildlife Collection.

**Chapter 6** *p. 154:* Norbert Wu. *p. 159:* Jack Jackson/Planet Earth Pictures. *p. 161:* (*top*) Norbert Wu; (*bottom*) Robert Arnold/Planet Earth Pictures. *p. 164:* Peter Scoones/Planet Earth Pictures. *p. 165:* Richard Herrmann/The Wildlife Collection. *p. 173:* After R. McN. Alexander, *Animals. p. 175:* Dr. Wagner. *p. 178:* Norbert Wu. *p. 179:* Peter David/Planet Earth Pictures. *p. 180:* Geoff Harwood/Planet Earth Pictures. *p. 182:* D. Parer and E. Parer-Cook/AUSCAPE International.

**Chapter 7** *p. 184:* Jean-Paul Ferrero/AUSCAPE International. *p. 187:* Edward Ross. *p. 188:* D. Parer and E. Parer-Cook/AUSCAPE International. *p. 189:* After R. McN. Alexander, *Carolina Biology Reader #164: Animal Movement*, Carolina Biological Supply Co., 1985. *p. 190:* Jim Brandenberg/Minden Pictures. *p. 195:* Francois Gohier/AUSCAPE International. *p. 196:* Norbert Wu. *p. 199: Scientific American* 252(1):96, 1985. *p. 201:* Charles McCutchen, National Institutes of Health. *p. 202:* Norbert Wu. *p. 203:* J. Gray, *Journal of Experimental Biology* 10:88–104, 1933. *p. 205:* Charles McCutchen, National Institutes of Health.

**Chapter 8** *p. 214:* Kathie Atkinson/AUSCAPE International. *p. 216:* M. Abbey/Photo Researchers. *p. 217:* M. Walker/NHPA. *p. 219:* G. Leedale/Biophoto Associates. *p. 220:* After T. P. Stossel, *American Scientist* 78:408–423, 1990. *p. 221:* Charles Brokaw, from T. Wu, C. Brokaw, and C. Brennen (Eds.), "Mechanisms of movement in flagella and cilia," *Swimming and Flying in Nature*, Plenum Press, 1975. *p. 224:* G. Leedale/Biophoto Associates. *p. 226:* Photo Researchers. *p. 229:* Aileen N. C. Morse. *p. 230:* Kathie Atkinson/AUSCAPE International. *p. 231:* Omikron/Photo Researchers. *p. 236:* John Kessler, University of Arizona.

**Epilog** *p. 238:* Stephen Dalton/NHPA.

# Index

Other books in the Scientific American Library Series

POWERS OF TEN
by Philip and Phylis Morrison and the Office of Charles and Ray Earnes

HUMAN DIVERSITY
by Richard Lewontin

THE DISCOVERY OF SUBATOMIC PARTICLES
by Steven Weinberg

FOSSILS AND THE HISTORY OF LIFE
by George Gaylord Simpson

ON SIZE AND LIFE
by Thomas A. McMahon and John Tyler Bonner

THE SECOND LAW
by P. W. Atkins

THE LIVING CELL, VOLUMES I AND II
by Christian de Duve

MATHEMATICS AND OPTIMAL FORM
by Stefan Hildebrandt and Anthony Tromba

FIRE
by John W. Lyons

SUN AND EARTH
by Herbert Friedman

ISLANDS
by H. William Menard

DRUGS AND THE BRAIN
by Solomon H. Snyder

THE TIMING OF BIOLOGICAL CLOCKS
by Arthur T. Winfree

EXTINCTION
by Steven M. Stanley

MOLECULES
by P. W. Atkins

EYE, BRAIN, AND VISION
by David H. Hubel

THE SCIENCE OF STRUCTURES AND MATERIALS
by J. E. Gordon

THE HONEY BEE
by James L. Gould and Carol Grant Gould

ANIMAL NAVIGATION
by Talbot H. Waterman

SLEEP
by J. Allan Hobson

FROM QUARKS TO THE COSMOS
by Leon M. Lederman and David N. Schramm

SEXUAL SELECTION
by James L. Gould and Carol Grant Gould

THE NEW ARCHAEOLOGY AND THE ANCIENT MAYA
by Jeremy A. Sabloff

A JOURNEY INTO GRAVITY AND SPACETIME
by John Archibald Wheeler

SIGNALS
by John R. Pierce and A. Michael Noll

BEYOND THE THIRD DIMENSION
by Thomas F. Banchoff

DISCOVERING ENZYMES
by David Dressler and Huntington Potter

THE SCIENCE OF WORDS
by George A. Miller

ATOMS, ELECTRONS, AND CHANGE
by P. W. Atkins

VIRUSES
by Arnold J. Levine

DIVERSITY AND THE TROPICAL RAIN FOREST
by John Terborgh

STARS
by James B. Kaler